T0134550

The Road to Servitization

Alessandro Annarelli · Cinzia Battistella · Fabio Nonino

The Road to Servitization

How Product Service Systems Can Disrupt Companies' Business Models

Alessandro Annarelli
Sapienza University of Rome
Rome, Italy

Cinzia Battistella
University of Udine
Udine, Italy

Fabio Nonino
Sapienza University of Rome
Rome, Italy

ISBN 978-3-030-12253-9 ISBN 978-3-030-12251-5 (eBook)
https://doi.org/10.1007/978-3-030-12251-5

Library of Congress Control Number: 2018968378

This Springer imprint is published by the registered company Springer Nature Switzerland AG
The registered company address is: Gewerbestrasse 11, 6330 Cham, Switzerland

To our beloved Pietro.

Cinzia and Fabio

To my family,
so proud of me,
even if they are still figuring out
what my research is about.

Alessandro

Preface

Numerous companies across industries are attempting to achieve competitive differentiation and generate higher customer value through servitization. This transformation from being physical product provider to service or product service system provider promises several advantages, such as increased revenue, stable income, offer differentiation and higher customer satisfaction. Common industrial examples of product service system include selling mobility instead of car or offering availability instead of a physical tool or guaranteeing performance or outcomes instead of a product with service contact. A key assumption related to product service system has to do with greater need to engage in value co-creation with customer by offering highly customized solutions to solve their problems. Although companies are increasingly recognizing the importance of product service system, many companies still struggle with the challenges of successfully selling (e.g., profiting) from servitization.

This book took a practitioner perspective towards advancing their understanding towards how to effectively undertake product service system transformation. Regardless if you are an executive from business-to-business or business-to-consumer environment, this book provides you with a much-needed overview on how to develop capabilities, introduce new organization practices, establish collaboration with value chain actors and most importantly transform your product-centric business model. Another advantage of this book is to be able to consolidate the fragmented literature on servitization and product service system. This successfully combines and integrates insights from product service system related concepts, such as industrial product service systems, hybrid product, service innovation, and service offerings. These streams of interrelated research provide missing pieces of the puzzles that the book successfully integrates. Finally, the book provide numerous real-life case examples (e.g. Xerox, IKEA, Rolls Royce and others) for establishing a strong link between theory and practice.

The book is organized into six chapters. First chapter focuses on the defining and conceptualizing product service system and presents challenges faced by companies as they undertake servitization transformation. Second chapter extends the provider view on how they engage in value co-creation with customers. Third chapter connects to the sustainability dimension of product service system. Explanation on how companies can achieve sustainability goals while still profiting from services are presented. Emerging trends such as circular economy, shared economy and digitalization are also discussed. Fourth chapter explains how companies achieve strategic advantages through product service system. This chapter connects to strategic management-related theoretical concepts, such as path dependency and strategic choices, and by doing so links product service system with established theoretical perspective. The fifth chapter takes a closer look at internal operational practices that need to be revised or updated in light of product service system strategy implementation. These practices relate to introduction of new strategic activities, implementation of diverse resources and utilization of key partnerships in the value chain. The final chapter brings all the previous discussions into the product service system business model framework. Without doubt, business model remains the key bottleneck to product service system implementation. This chapter exemplifies how to work with different elements of a business model as companies move from product-oriented to use-oriented to result-oriented business models. Thus, the book provides a rich and in-depth perspective towards how product service system is disrupting traditional business models.

I have been working on the topic of servitization and product service system for more than 10 years. This work has been published in 100 plus academic publications and build on data from numerous leading companies, like ABB, Ericsson, Volvo, Sandvik, SCA, Komatsu Forest, Scania, Metso and others. A common challenge for many of these leading companies remains how to move away from product centric business model to more innovative product service system business models. The challenges are numerous but only through understanding them, it is possible to effectively tackle them. I would recommend the current book to both young academics and practitioners as they began their journey towards servitization and product service system. Moreover, I would like to congratulate authors for an excellent job with writing this book and wish them all the success!

Luleå, Sweden

Vinit Parida
Luleå University of Technology (Sweden)
Chair Professor of Entrepreneurship and Innovation
April 2019

Contents

1 What is a Product Service System? 1
1.1 Product Service System: Three Words, Multiple Implications 1
1.1.1 Classifications of Product Service Systems 3
1.1.2 The Three Literature Perspectives 6
1.1.3 From Product Service System to Servitization: Different Terms for the Same Concept? 9
1.2 The Challenging Transition of Servitization: Integrating and Bundling Products and Services 16
1.2.1 Degree of Servitization in the Product-Service Continuum 20
1.2.2 Drivers of Servitization 21
1.2.3 Benefits and Barriers of Servitization 22
1.2.4 The Service Paradox 24
1.2.5 The New Role of Client 25
References 26

2 The New Role of Client: From Ownership to Value Co-creation . . . 31
2.1 Servitization as a New Value Proposition 31
2.1.1 Service Offering 32
2.1.2 Customer Value 37
2.1.3 Value Co-creation 39
2.1.4 Accessing the Value: From Ownership Towards Use 41
2.2 The Key Issue in the Customer Management 44
2.2.1 Improved Relationships with Customers 44
2.2.2 Customer Interaction 45
2.2.3 Information Sharing 46
2.2.4 Sales Channels' Effect in Value Communication 47
2.2.5 Different Contractual Models 49
References 52

3 Product Service Systems' Competitive Markets 55
3.1 Contemporary Social and Economic Context 55
3.1.1 Sustainability 55
3.1.2 Sharing Economy and Collaborative Consumption 59
3.1.3 Circular Economy 64
3.2 Product Service Systems in B2B and B2C Markets 67
3.3 Product Service System in the Traditional Manufacturing Industries 70
3.4 Product Service System in the Sustainability-Driven Industries . . . 74
3.5 Product Service System in the New Digital-Driven Industries 85
References 93

4 How to Trigger the Strategic Advantage of Product Service Systems 95
4.1 Translating PSS into Competitive Strategy 95
4.1.1 The Process of Strategy Formulation 96
4.1.2 The Role of Path Dependence in the Strategy Formulation 98
4.2 Traditional Strategies Driven by PSS 100
4.2.1 Cost Leadership 102
4.2.2 Product and Service Differentiation 104
4.2.3 Niche Strategy 107
4.3 Drivers for Competitive Advantage of Product Service System . . . 109
4.3.1 Closing the Loop: Seizing the Opportunities of Circular Economy 111
4.3.2 Using Rather Than Buying: Pursuing the Sharing 113
4.3.3 The Need for Differentiation and Innovation 117
4.3.4 New Market Segmentation 120
4.3.5 Inimitability and Protection from Replicability 121
4.3.6 Loyalty as a Measure of Success 122
4.4 Evaluating Sustainability of PSS Competitive Advantage 123
4.4.1 Analysing PSS Risks 123
4.4.2 Assessing PSS Sustainability 126
4.4.3 Estimating the Strategic Value of Servitization 133
References 138

5 Translating PSS Strategy into Operations 143
5.1 The New Role of Operations and Service Strategy 143
5.2 Key Activities for Product Service System Design 144
5.2.1 Product and Service Design 145
5.2.2 Product and Service Configuration Support 152
5.2.3 Product and Service Delivery 153
5.2.4 Functional Integration 155

5.3 Key Resources in PSS Implementation 155
5.3.1 ICT and Monitoring Technologies 156
5.3.2 Installed Base Information 159
5.3.3 Human Resources 159
5.3.4 Financial Resources 161
5.4 Key Partners in PSS Implementation 161
5.4.1 Network 163
5.4.2 Supplier Relationship 168
References 170

6 How Product Service System Can Disrupt Companies' Business Model 175
6.1 Decomposing PSS Business Models 175
6.1.1 Business Model Canvas 176
6.1.2 Business Model Innovation Process 179
6.2 Key Elements of Product Service System Business Models 180
6.3 Unveiling the Key Elements of PSS-Based Business Models 188
References 203

Glossary 207

Index 213

Introduction

Product service systems (PSSs) are business models based on the offering of a mix of both products and services. New solutions are emerging as innovative means to enable collaborative consumption of both products and services following trends of sustainability, circular economy and dematerialization.

A business strategy based on product service system establishes a value proposition focused on final users' needs rather than on the product, allowing for an easier design of a need-fulfilment system with radically lower impacts, in terms of environmental and social benefits.

Proper knowledge of the overall product service system phenomenon is essential to fully understand new competitive forces, related to newest trends like sharing economy and circular economy. In the contemporary economic and social context, product design and manufacturing can no longer be the only source of competitive advantage and differentiation: product service integrated solutions bring innovation potential adding value to the total offering. This could be also the simple case of extra services added to the product offering with the aim of prolonging product life cycle and utility through time (for a more sustainable performance), while providing to customers a more satisfactory experience worthy of extra revenue. Product service system unique capability of addressing all three pillars of sustainability (economic, environmental and social) makes it a win-win strategy for companies, securing competitive capabilities and resources while ensuring an extensive exploitation of manufacturing means.

Understanding how the transformation of business models happens and how to design and to manage the processes through which successful businesses and managers develop these new business models are key topics attracting the attention of a rising number of practitioners and scholars. In recent years, in business model studies, some interesting topics have emerged proposing feasible ways to business innovation linked to sustainability concerns. And these topics are closely linked to product service system. While research about PSS has been well established for more than 20 years, there is still growing attention and the need to have a deep understanding of successful business models.

The considerable number of elements involved in a PSS can potentially bring a considerable number of obstacles and barriers, deriving from complex issues and difficulties linked to these many elements and their interaction. These issues need the analysis of the strategic value of PSS and its ability to create a disruptive and sustainable value proposition. Understanding the actual role of PSS as a value proposition is a key aspect and, to achieve this purpose, is fundamental to evaluate the actual results obtained by firms, thanks to PSS implementation.

Despite the variety of possibilities concerning PSS and its implementation, which involve different degrees of change radicalness, there is still a lack of effective guidelines to support and guide companies and managers in its effective adoption. In other words, the majority of companies still need practical and theoretical established knowledge to deepen their competences on how PSS's could contribute in reaching their strategic goals, analysing its strategic value and understanding how to implement it.

Objective of the Book

Aim of this text is to provide a guidance for practitioners and scholars throughout the topics of servitization and product service system. Our book begins by describing current state of the art of product service system and providing a framework to categorize knowledge about the topic; then it presents the evolution and spread of the PSS models across industries, and presents current and most relevant applications; then it highlights its real value for strategy and management, operations and sustainability. Furthermore, it suggests how to enhance its design and implementation through an overview of relevant elements to be considered when planning and designing a product service system-related offering.

Underlying questions to which the book responds are as follows:

- *How are PSSs related to servitization, circular economy and collaborative consumption?*
- *How does the context change because of a PSS?*
- *How can companies obtain a competitive advantage from a PSS?*
- *Which are the key elements of PSS business models?*
- *How should companies implement PSSs?*

Structure and Contents of the Book

The book contains six chapters visually represented in the figure below and contains 1 exhibited with a qualitative study, and 11 boxes reporting 14 case studies.

The overall framework depicted in figure is both a representation of fundamental elements that managers should consider when moving towards servitization and a

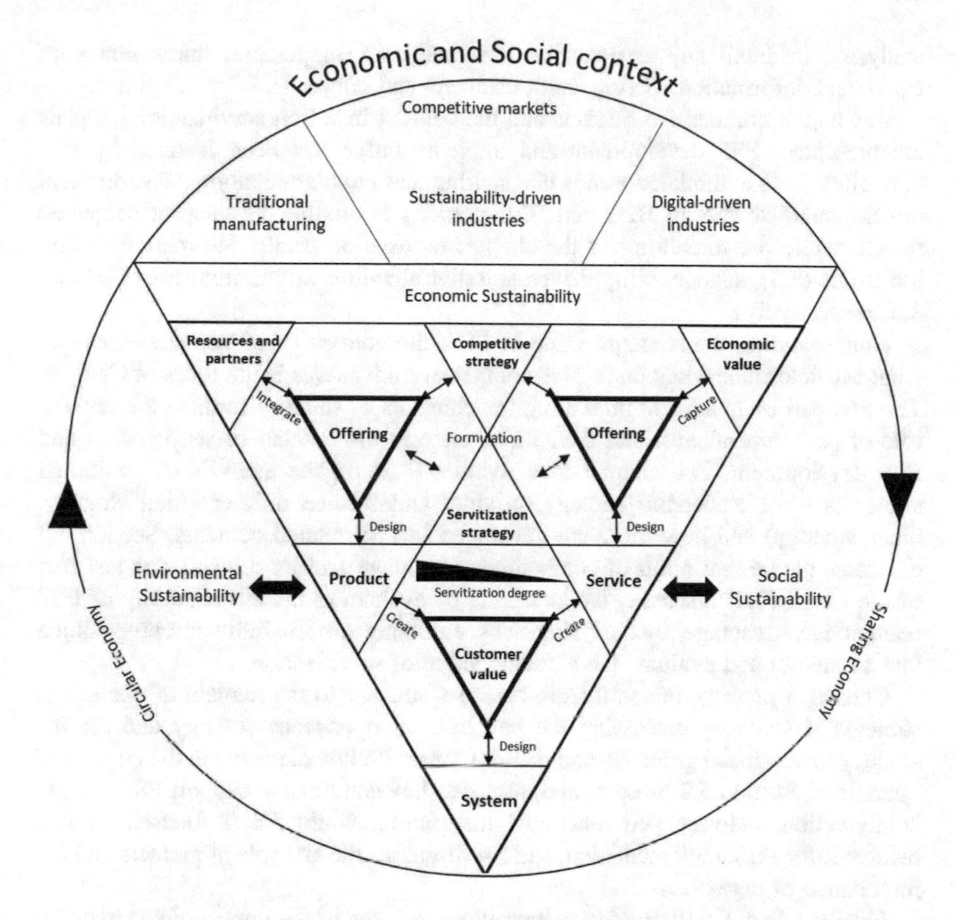

business models based on PSS and both the representations of the book structure. It guides readers to comprehend actual and potential applications of product service systems, thanks to scientific evidences and exemplary cases that led companies to innovative and winning business models. Moreover, it guides reader to a correct selection and implementation of PSS in different industrial environments.

Chapter 1 introduces the concepts of servitization and product service system; the first part of the chapter (Sect. 1.1) presents classifications of PSSs and deals with the origins of the terms and definition adopted. The second part (Sect. 1.2) is aimed at presenting and discussing the key issue of integrating products and services in the transition toward servitization: the sub-paragraphs will focus on key concepts like the degree of servitization, benefits, barriers and drivers behind servitization, the service paradox, and will then introduce the key topic of the following chapter, namely, the new central role played by customers in a servitized context.

Chapter 2 starts by presenting servitization as a new value proposition (2.1), by studying in detail the key concepts of service offering, customer value, value co-creation and value access. Customer involvement is discussed in 2.2 by analysing in detail key aspects like relationships management, interaction with customers, information sharing, sales channels and contracts.

In Chap. 3 competitive markets and the context in which servitization happens are presented. PSS development and implementation has been fostered by sustainability (3.1) and related trends like sharing and circular economy. The different development of PSS in B2C and B2B markets is another key aspect deepened in 3.2, while the remaining of the chapter focuses on traditional manufacturing industries (3.3), sustainability-driven and digitalization-driven industries (3.4 and 3.5, respectively).

Understanding the strategic value of PSS, the content of related strategies and what can determine a key distinctive competitive advantage is the focus of Chap. 4. The first part (4.1) aims at presenting the concepts of strategy formulation and the role of path dependence, and their importance in the context of servitization and PSS development. The chapter then focuses (4.2) on the analysis of traditional strategies (cost leadership strategy, product and service differentiation strategy, niche strategy) and how these are developed in PSS-related contexts. Section 4.3 discusses the central point of competitive advantage and its drivers. The last part of the Chap. (4.4) discusses the key issue of evaluating the sustainability of PSS competitive advantage, by analysing risks, assessing sustainability and proposing a tool to predict and evaluate the strategic value of servitization.

Chapter 5 presents the shift from business strategy to the domain of operations strategy: it starts by discussing the new role of operations strategy and service strategy in servitized contexts, and then it focuses on key elements in the context of operations. Section 5.2 presents and discusses key activities to support PSS design, configuration, delivery and functional integration, while 5.3 is focused on key resources for PSS implementation, and 5.4 discusses the key role of partners and the importance of networks.

Finally, Chap. 6 will provides a thorough overview of the impact of servitization on the overall business model of companies: this final part of the book is intended to be read as a summary of previous chapters, with an all-encompassing perspective on the topic in the managerial context. The first part (6.1) introduces two models that can be adopted to analyse/decompose a business model, while 6.2 presents key distinctive elements behind PSS-related business models. The last part of Chap. (6.3) presents case studies that exemplify the presence of servitization elements in related business models.

Our book provides empirical evidences throughout the chapters for most of the investigated subjects either from scientific theories and practical and industrial experiences. It also includes illustrations, charts and tables to effectively communicate with readers. A final note on case studies. Four cases are original and proposed for first time, while the 10 remaining ones have been carefully selected from literature with the objective of providing tangible examples of the strategic and operative frameworks and of the best practices presented in the book.

Chapter 1
What is a Product Service System?

This chapter deepens the concepts of servitization and product service system, presenting main insights emerging from both academic studies and practitioners' experiences.

1.1 Product Service System: Three Words, Multiple Implications

The *Servitization* of business is a from an exclusive focus on products or an exclusive focus on services towards integrated systems or bundles of products and services, with services playing a relevant role.

Even if the term servitization is the first to appear at the end of 80s (e.g. Vandermerwe and Rada 1988), the topic of servitization strategy witnessed a real development only at the end of 90s, when the term *Product Service System* (PSS) appeared for the first time. Its original definition is "a Product Service System is a marketable set of products and services capable of jointly fulfilling a user's need. The PS system is provided either by a single company or by an alliance of companies. It can enclose products (or just one) plus additional services. It can enclose a service plus an additional product. And product and service can be equally important for the function fulfilment" (Goedkoop et al. 1999, p. 18).

This definition of PSS highlights three main elements, which are the product, the service and the concept of system.

- *Product*: A tangible element, a good conceived, designed and manufactured in order to be sold with the aim to answer to customers' demand for satisfying their needs.
- *Service*: An action or activity performed to help, to do a work, to complete a task for others, specifically customers. It has a value, and it is exchanged/performed on a commercial basis. A service is usually defined by four key characteristics which are intangibility (cannot be inventoried, display, communicated), heterogeneity

A. Annarelli et al., *The Road to Servitization*,
https://doi.org/10.1007/978-3-030-12251-5_1

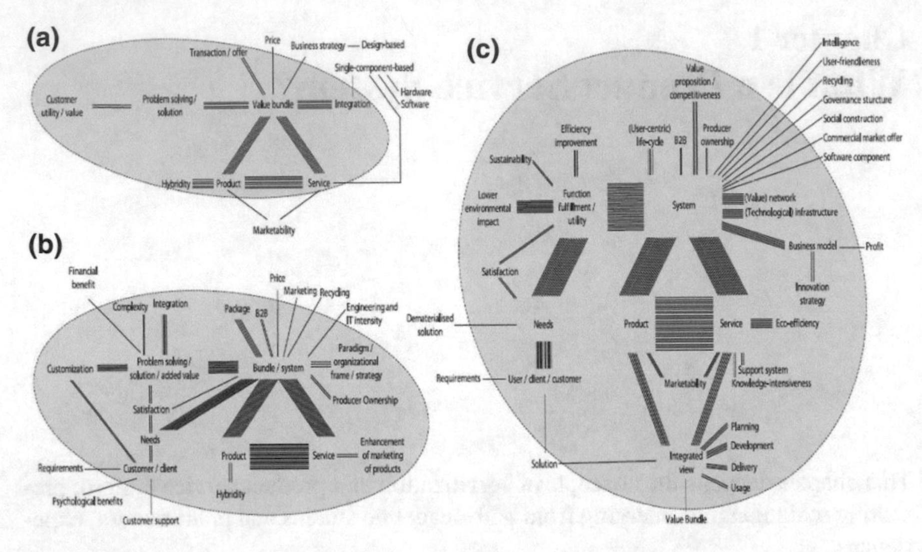

Fig. 1.1 Maps of main concepts in PSS definitions (Boehm and Thomas 2013)

(service delivery and customer satisfaction depend on the behaviour of service providers and customers), inseparability (customers participate in and affect the creation and delivery of a service) and perishability (services cannot be stored, returned, resold).

- *System*: A set of interrelated elements, comprising both the elements and their relationships.

The combination of these three elements brings towards the definition of product service system, which is a set of products and services combined to jointly achieve the fulfilment of customers' need.

PSS concept has been then strictly connected to very different concepts, ranging from "system" and "integrated view" to "business model" and "strategy" to "ownership" and "customization". A complete map of the most recurring concepts in definitions has been drawn by Boehm and Thomas (2013) and is represented in Fig. 1.1 while it gives an idea of how many elements can occur inside the concepts of PSS and servitization. The three areas represented concern, respectively, definitions in the field of (a) Information System, (b) Business Management, and (c) Engineering and Design. Some examples of the most recurring elements are, for instance, business model, lower environmental impact, eco-efficiency, integrated view, problem-solving, customization, hybridity and considering the most predictable ones: product, service, system, bundle, users, needs, fulfilment and solution.

1.1.1 Classifications of Product Service Systems

PSSs can be categorized in many ways. The most widely accepted and reported categorization of PSS is based on the object of the contract and the shift in ownership (Tukker 2004).

The classification scheme in Fig. 1.2 shows the three main categories of PSS (product-oriented, use-oriented and result-oriented) and how they are located along the product–service continuum.

In the first category, named *Product-Oriented PSS*, the focus is on the product, i.e. the PSS is oriented towards product selling, with extra services added to the offering. In this case, there are two possible service configurations:

- Product-related service, where services offered are related to the usage phase of the product, e.g. preventive maintenance or spare parts provision;
- Advice and consultancy services, like advices for logistic optimization, education and training on product usage, or either financial services.

With the *Use-Oriented PSS* the attention shifts from selling the product to access to usage: the same product is accessible to different customers in a limited time span according to different forms of renting and/or sharing. Indeed, services in this case might be differentiated in:

- Product lease, where customer pays for the access and usage of the product for a considerable amount of time obtaining an exclusive (individual and unlimited) use of the product for the time of the subscription;
- Product renting or sharing, where the product can be sequentially rented and used by different customers, even during the same day; the main difference between the forms of renting and sharing lies in the dynamicity of the service offered and of the renting formula;
- Product pooling, where a simultaneous use of the product from different customers at a time is allowed.

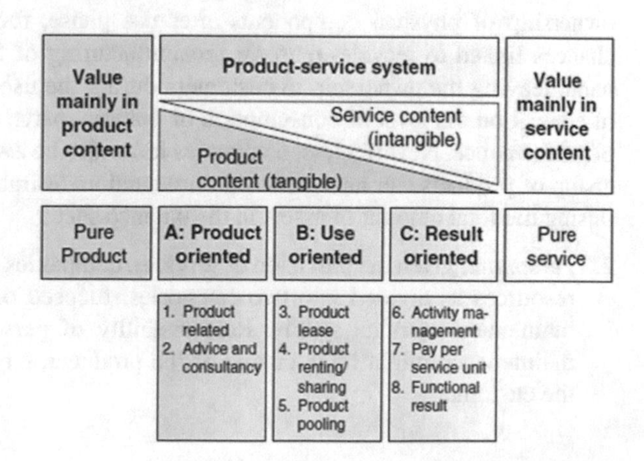

Fig. 1.2 Categorization of PSSs (Tukker 2004)

The third category proposed is focused towards the final result provided by the product, as the name *Result-Oriented PSS* highlights. In this case, the producer/provider maintains the ownership of the product, and we can therefore state that the usage of the product itself is in some way outsourced by the customer. Services in this category can be distinguished in the following:

- *Activity management* and/or *activity outsourcing*, where the agreement between provider and customers concerns the outsourcing of an activity;
- *Pay per service unit*, which presents different similar characteristics with use-oriented services but, in this case, the customer directly pays for elementary units of output provided;
- *Functional result*, which can be considered the most radical example of servitization/PSS with customer and provider agreeing only on the result to be delivered and usually no significant constraints on how to deliver the result.

The more radical the degree of servitization is, the more "abstract" the customer's need becomes. These characteristics bring a certain freedom of action that results in a dualism opportunity/threat since, on one hand, there is a concrete freedom on how to deliver a result but, on the other hand, there are hidden difficulties in translating concretely the abstract needs of customers.

Five key elements can be used as lenses to study and understand a PSS business model and related offerings (Lay et al. 2009). These are as follows:

1. *Ownership* of the product/physical component, which can be divided on a temporal basis into:
   - *Ownership during use*: maintained by the producer or the supplier, an intermediate actor like a leasing bank or an operating joint venture, or transferred to the customer;
   - *Ownership after use*: maintained by the supplier or either returned to the producer for the reuse/remanufacture of physical components.

Companies should carefully manage the state of ownership, since maintaining ownership of physical components after use phase, for instance, might disclose chances linked to recycle, reuse or remanufacturing of components. On the other hand, leaving the ownership to customers during the use phase could imply a lack of control on the level of consumption of material parts, causing an unwanted lack of performance. Nevertheless, companies must also be aware that maintaining ownership of products can have non-negligible and undesirable economic impacts, e.g. raising the total amount of assets in the balance sheet.

2. *Personnel*: given the provision of services, companies must consider their human resources as divided into two categories, focused on either manufacturing or maintenance/service supply. Responsibility of personnel for production and maintenance might be in charge of the producer, an operating joint venture or the customer.

Characteristic Features		Options			
Ownership	during phase of use	Equipment producer	Leasing bank	Operating joint venture	Customer
	after phase of use	Equipment producer	Leasing bank	Operating joint venture	Customer
Personnel	Manufacturing	Equipment producer	Operating joint venture	Customer	
	Maintenance	Equipment producer	Operating joint venture	Customer	
Location of operation		Equipment producer's establishment	Establishment "fence to fence" to the customer	Customer's establishment	
Single / multiple customer operation		In parallel operation for multiple customers	Operation for a single customer		
Payment model		pay per unit	pay for availability	fixed rate	pay for equipment

Fig. 1.3 Morphological box for understanding a PSS (Lay et al. 2009)

3. *Location of operations* can be either at the producer's or customer's establishments, but the model considers also a third way, where the supplier establishes a "fence-to-fence supply park" to work in closer cooperation with the customer.
4. *The number of customers served* at the same time can simply vary from one customer at a time to multiple customers served in parallel. This characteristic is strongly dependent upon the amount of product and service components involved in the offering, and companies can have a limited degree of freedom on this aspect.
5. *Payment model* is the last parameter considered in the framework. Customers might pay for the equipment/utilities as in a traditional purchase model, or other ways of flexible payment can be implemented like, for instance, payment for actual utilization time (pay for availability), for units effectively produced or payment for time unit (fixed rate).

The implementation of a PSS carries to a plethora of suitable options: on the right side of Fig. 1.3, all options related to a traditional purchase-only solution can be found, where, for instance, the supplier sells a machinery to a customer in charge of operating, maintaining and then managing its dismissal at the end of the useful life cycle. On the other extreme, to the whole left of Fig. 1.3, can be retrieved all options related to a complete outsourcing offering, where the producer maintains the ownership of the equipment and all responsibilities linked to its functioning and to the management of personnel for manufacturing and maintenance tasks. This framework will be detailed in the key aspect of ownership also in paragraph 2.1.4.

1.1.2 The Three Literature Perspectives

Literature on servitization and PSS benefitted from contributions coming from different research fields. In fact, since its origins, PSS attracted the interest of design researchers because of its nature of socio-technical system. This term first appeared in 1960 and was coined by Emery and Trist "to describe systems that involve a complex interaction between humans, machines and the environmental aspects of the work system. The corollary of this definition is that all of these factors—people, machines and context—need to be considered when developing such systems using Socio Technical Systems Design methods" (Baxter and Sommerville 2011).

PSS is an interdisciplinary field because it presents interesting and challenging characteristics for many researchers from different research areas. Business management mostly investigates the bundling of products and services from a marketing perspective while, in the Engineering & Design field, the focus is on designing, developing and delivering the PSS to the final user. Finally there is a developing interest from the ICT and Information Systems disciplines because of the increasingly close relationship between PSS and technology.

These different sources of knowledge originated streams of research on the topic with different points of view emerging. Indeed, literature on servitization can be divided into contributions focusing on design of the offer and/or business model, or specific parts within them, others focusing on environmental (and related social) issues and the overall benefit coming from servitized models, while the third group focused closely on economic insights.

The Design Perspective

A considerable amount of academic production on the topic of PSS comes from the Engineering and Design discipline, mainly focused on the design, building and operating conditions of products. Inside this field many researches contributed to the definition of PSS and also of its synonyms like "functional sale", "functional products", "total care products", "extended products" and "servicification". This group contains also publications that dealt with new service development and service engineering topics.

In this stream, there is the highest concentration of contribution in literature (Annarelli et al. 2016). This disproportion clearly shows that literature lacks a deep insight into the evaluation of PSS's economic and environmental/social impact.

In this group, there are seminal works on PSS, the most cited in literature, like those by Vandermerwe and Rada (1988), Roy (2000), Mont (2002), Manzini and Vezzoli (2003), Oliva and Kallenberg (2003), Baines et al. (2007) and Meier et al. (2010). All these papers mainly provide a general overview of PSS characteristics and potentials, summarizing them in theoretical frameworks with a clear theory building research purpose. However, the main sub-field of research that clearly emerges in this group is *PSS design and development*: also if many papers are focused on PSS strategic aspects, their final aim is providing guidelines, tools and/or methodologies for an effective PSS design process.

The prevalence of the “design issue” is the clearest difference of this group with the others, where design was a side topic or not considered at all. Then, it is possible to conclude that the development/design of PSS contrasts with economic/social/environmental analysis. Moreover, behind these two tendencies, one towards design and the other one towards economic/social/environmental analysis, the presence of different viewpoints on PSS is quite evident, also on a time perspective: the first tendency (design) looks at the future and at what can be done; the second one (economic/social/environmental analysis) looks at the past and at what has already been done. This consideration can help in identifying a promising future research need, linked to the aim of mixing and promoting the coexistence of these two tendencies, especially considering the potential benefits of implementing the the results of economic/social/environmental analyses into designs of future PSS.

An important aspect that received little attention is the nature of PSS as a socio-technical system. Roy (2000) first acknowledged this characteristic of PSS, stating that it could provide essential end-use functions and resulting in better environmental and consumption performance rather than traditional products sold. After this work, scholars from design fields mainly focused on design methodologies and/or tools, and the concept of PSS as a socio-technical system has been reconsidered in recent years (Ceschin 2014; Rivas-Hermann et al. 2015).

In particular, Ceschin (2014) starts from the premise that there is a need for a deep redefinition of consumption and production habits to ensure a successful adoption of sustainable PSS, acknowledging that PSS does not simply constitute a new offering, but can be viewed as a social innovation and a large-scale socio-technical change.

This radical change must involve the identification of the most appropriate “strategies and pathways to favour and hasten the introduction and scaling-up” of sustainable PSSs. That is why the author recognized that “the introduction of radical innovations requires the creation of partially protected socio-technical experiments. [...] Protection allows incubation and maturation of radical socio-technical configurations by partly shielding them from the mainstream market selection environment”.

The Environmental Perspective

Contributions grouped in this area have all in common the interest towards the environmental potential brought by PSS and, even in different ways, all papers in this group provide a tool/framework to analyse the contribution of PSS from the point of view of environmental burden, e.g. by means of analyses concerning reduction in waste production, reduced amount of production cycles or reduction of inputs in the production process.

A smaller percentage of literature deals with the environmental/social analysis of PSS, but, starting from 2012 to 2015, there has been a rise in the number of environmental/social analyses.

The greatest majority of studies has a clear focus on sustainability, while few other papers are focused, respectively, on strategy, production, and on design. Therefore, the analysis of environmental/social impact can be linked to sustainability and strategic aspects.

Going into detail, what first emerges is that this analysis is much more qualitative in nature than economic perspective. Dewberry et al. (2013), dealing with PSS design and development process, provide a framework for "Home life cycle" analysis considering the four different phases of (1) specification and sale, (2) use, (3) disposal, (4) re-sale and use. Halme et al. (2004), in their work on the environmental/social assessment of household services, provide an "operazionatilization of sustainability indicators", using a scale to assess impacts of PSS change. Maxwell and van der Vorst (2003) describe the features of a method for sustainable product and service development providing an overview of the overall process and analysing how it can be incorporated into an organization's processes and systems. Briceno and Stagl (2006) investigate the PSS social effects of local exchange trading schemes; by surveying organizers and participants, authors provide a clear overview on PSS's effectiveness in social terms. Evans et al. (2007) provide a very useful tool for assessing and representing "Environmental improvements for SME employee solution". In a more technical way, Tasaki et al. (2006) provide a quantitative method to assess material use and consumption level in "lease/reuse systems of electrical and electronic equipment". Firnkorn and Müller (2011), analysing car sharing systems, deploy the system into processes, parameters and effects in order to assess the overall environmental impact.

The Economic Perspective

Papers in this area mainly focus on the economic and financial impact of PSS: indeed, understanding the actual contribution of PSS to the overall business dimension, e.g. in terms of revenue gains or costs reduction, is the main aim of works gathered in this sub-field.

Papers in this group provide an analysis of PSS economic potential, mainly in quantitative terms. From a time-trend perspective, papers belonging to this group show a rising trend starting from 2009 to 2012.

The great majority of papers is focused on the strategic value of PSS and economic analysis/evaluation most of times derives from strategic considerations. None of the papers belonging to this group has a side focus on sustainability.

The economic analysis of PSS is carried out in several ways with different methodologies. Azarenko et al. (2009) propose a cash-flow analysis for a machine tools provider forecasting, for the following 20 years, the expected economic benefits of PSS and using this analysis to compare the product-oriented, use-oriented and result-oriented categories, in terms of monetary results. Similarly, other authors (Nishino et al., 2012; Kreye et al., 2014) mainly focus on cost estimation trying to evaluate the transition to PSS in a meaningful quantitative way.

Richter et al. (2010) provide an economic analysis in order to appraise in quantitative terms the evolution of business models when employing PSS: the analysis is performed only for the use-oriented category with the aim of estimating changes in costs, revenues and profits comparing the servitization alternative with cost-plus and fixed-price contracts showing that the PSS is a win-win situation for customer and supplier.

Neely (2009) compares firms on the basis of their sizes and focus (purely manufactured vs. servitized organizations) obtaining interesting results: servitized firms tend to generate higher revenues, but lower profits compared to pure manufacturing firms, and this is true for larger firms; for organizations with less than 3000 employees, this finding is completely inverted. This is called the *paradox of servitization* (Neely 2009). Finally, Friebe et al. (2013) explore low-income markets in the context of solar home systems evaluating the economic potentials of PSS. Komoto et al. (2012) show how the economic analysis of PSS can be implemented in the design phase, improving the design process and overall performance.

1.1.3 From Product Service System to Servitization: Different Terms for the Same Concept?

During the years, several terms have been created to indicate the same concept:

- *Industrial Product–Service System*, which is "characterized by the integrated planning, development and use of product and service shares in B2B applications and represents a knowledge intensive socio-technical system" (Meier et al. 2010);
- *Servicification*, "the increase in use, produce, and sale of services" (European Commission 2014);
- *Post Mass Production Paradigm*: the aim is that of decoupling economic growth from resource/energy consumption and waste creation. The aim is the transition from quantitative sufficiency to qualitative satisfaction, by expanding services range for manufacturing companies (Tomiyama 1997);
- *Functional Sale*, in which the company decides how to fulfil the function bought by the customer with functional products, total care products and integrated solutions (Davies 2004);
- *Hybrid Product*, *Hybrid Value Bundles* and *Hybrid Value Creation*, which are integrated offerings of products and services designed to meet specific customer demands and generate additional value.

In order to shed light on differences and similarities among the plurality of terms adopted, we can address the ontological and epistemological views behind these terms, as retrieved in literature (see exhibit "From Servitization To Product Service System: an ontological and epistemological view").

Exhibit—From Product Service System to Servitization: An Ontological and Epistemological View

Ontology is based on two different assumptions:

- Realist/Objectivist assumption: phenomena such as "organizations" exist "out there" independently from our perceptual or cognitive structure and attempts to know.

- Idealist/Subjectivist assumption: a social reality is a creation or projection of our consciousness and cognition

Epistemology is based on the way we construct reality and we give meaning to external events, so it is possible to operate a distinction between:

- Epistemological objectivity: we experience the world only through direct sensory experience, so the focus is exclusively on facts; the "true" is based on the statistics indicating performance and quality data, figures and graphics of finance and accounting, strategic documents reporting business plans, forecasts of performance derived from statistics.
- Epistemological subjectivity: we comprehend the social world through the meaning people give to their world; in an organizational context the focus is on social processes lying behind the production of documents, which motivates people in going on with their work/activity.

By means of these theoretical lenses, it is possible to provide a meaningful analysis of the term product service systems and its synonyms in the figure below.

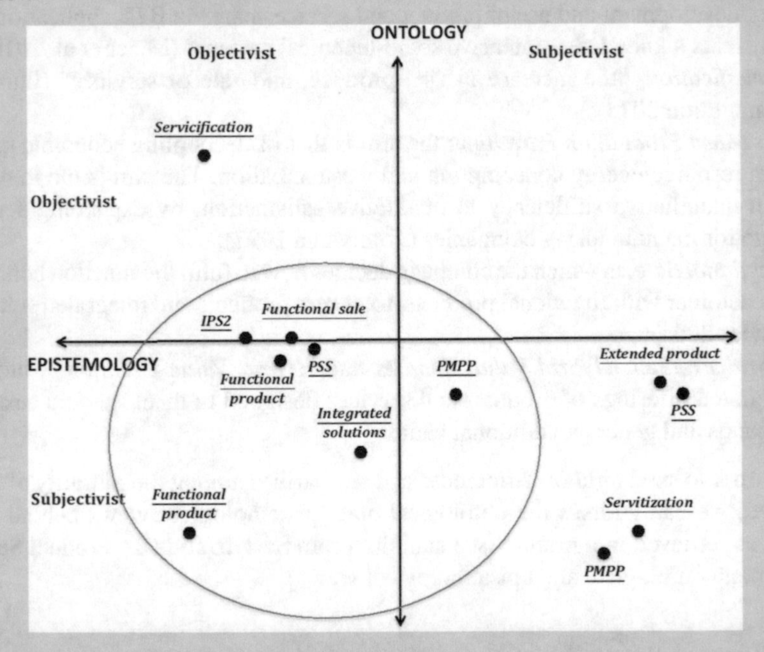

Streams of thought connected to different ontological and epistemological views.

The work from Goedkoop et al. (1999), which first introduced the term PSS, has a clear idealist/subjectivist perception of ontology, as emerges, for instance, from the following extract: "Economic growth is linked to perceived value creation and not necessarily to material or product streams circulating in the economy". On the other hand, the epistemological view presents both subjectivist and objectivist elements, with the first ones mainly linked to social dimension of sustainability and emerging from reflections on consumption schemes and how they are affected by PSS offerings. At the same time, there are elements of epistemological objectivity on the economic side with a clear need (highlighted throughout the entire paper) to give an economic assessment and evaluation of PSS possibilities. This can bring to classify the paper by Goedkoop et al. (1999) as an Interpretivist paper.

Product Service System—PSS

In the paper of Mont (2002), there is an ontological realist/objectivist perception, with the author stressing the objectivist and quantifiable nature of service industry/service economy. Like in the work from Goedkoop et al. (1999), the epistemological view can be considered as halfway between objectivism and subjectivism. In this case, the objectivist elements are focused on the environmental/social aspects of sustainability by stressing the need to give a clear evaluation of dematerialization and its effects on material flows, so as to assess savings in materials burden, reduction of environmental impact, consumption efficiency and production efficiency. Subjectivist elements of epistemology emerge from reflections on the nature of products' added value (in traditional offerings compared to PSS). The paper by Mont (2002) can be classified as a Pragmatist paper.

Servitization—Servicification

The work by Vandermerwe and Rada (1988), which first introduced the term Servitization, is a clear postmodernist work: there are strong elements of ontological subjectivity, mainly related to the definition of servitization and to reflections on nature of services and their "new" role. As for ontology, there is a clear epistemological subjectivist position emerging from paper's methodology: indeed, this work is based on interviews with senior executives of both services and manufacturing companies with the aim of emphasizing "the growing importance of services in corporate strategy."

From the analysis of the work by Lodefalk (2010), which introduced the term servicification, it clearly emerges the strong objectivist nature of this paper, from both the ontological and epistemological perspectives. That is why this paper can be classified as positivist. The author states that country and firm boundaries are less relevant, and every organization can choose to vertically integrate or specialize, at home or abroad, outlining four different strategies. This close categorization reveals a tendency to classify and describe through quantitative/discrete data the reality, this is a typical tendency of an ontological objectivist perspective. On the epistemological aspect, the paper is focused on analysing the servicification in the context of an industrialized country using micro-level data; the work is geared towards a description/classification of services encountered in the research activity, through quantitative measures.

Industrial Product Service System—IPS2

Meier et al. (2010) introduced the term Industrial Product–Service System (IPS2) using a Pragmatist view. The paper starts with a significant statement: "The world is changing. Industrialised countries are subject to a structural change toward a service society". To support this thesis, the authors provide some empirical evidences: United States recorded a percentage GDP from service in 2005 of 76%, Germany of 70% and Japan of 69%, revealing an objectivist approach in an ontological view. In this work, the B2B servitized environment is presented and described through a series of subjectivist considerations, like "Industrial Product Service System business demands a paradigm shifts towards selling functionality instead of selling products", together with statistical data and case studies clearly depicting the situation. The paper is focused through the provision of a framework for industrial product service system business modelling, a framework presenting an interesting mix of quantitative and qualitative considerations, positioning this work halfway between subjectivist and objectivist epistemology.

Post Mass Production Paradigm—PMPP

Tomiyama (1997) introduced the Post Mass Production Paradigm (PMPP). This paper can be categorized as a postmodernist paper with a well-declared tendency to subjectivist perception of ontological and epistemological aspects of research. The paper starts considering the mass production and mass consumption paradigms as "modern evils" because of some aspects connected to market saturation and natural resources uncontrolled usage. The author states: "Modern evils arise when technological advances encounter limitations concerning natural, social and human resources". This problem has been addressed by reconsidering the current (in 1997) paradigms of economic activity by reducing the volumes of production and consumption by balancing them with "natural, social and human constraints" to reach an "adequate, manageable size". The author suggests that this can be done through the Post Mass Production Paradigm by "decoupling economic growth from resource/energy consumption and waste creation". The aim is the transition from quantitative sufficiency to qualitative satisfaction showing a strong ontological-subjectivist position. On the other hand, the transition to this new paradigm is explained through a series of points focusing on qualitative characteristics of services range expansion for manufacturing companies in order to meet the objectives of sustainable production and consumption; all analyses in this paper are mainly at a qualitative level, which is clearly a proof of epistemological subjectivist thought.

The paper from Umeda et al. (2000), also dealing with Post Mass Production Paradigm (PMPP), opens by re-proposing the same considerations of Tomi-yama (1997) about mass production and mass consumption and the need of re-configuring current (in 2000) business models in the light of the Post Mass Production Paradigm. However, the paper also highlights the need for a structural change of product/service/life cycle design which plays a central role in reaching the goal of environmental and social sustainability. Like the previous paper (Tomiyama 1997), this work presents a strong subjectivist ontological perspective towards the topic of sustainability in production and consumption, but it has also some objectivist points. On the epistemological side, the paper proposes a "methodology for the life-cycle design to establish sustainable closed-loop product life cycles. [...] The simulation system evaluates product life cycle from an integrated view of environmental consciousness and economic profitability and optimizes the life cycles. This paper also discusses feasibility and advantages of this simulation system by illustrating a case study." The considerations above can bring to classify this paper as a mean way between Interpretivism and Conventionalism.

Functional Sale

The paper authored by Sundin and Bras (2005) adopts the term functional sale and can be classified as a positivist paper with also some Pragmatist characteristics. This work starts with environmental considerations about unsustainable patterns of production and consumption, highlighting the need for a more sustainable development based on closed-loop material flows. This goal can be reached "by a larger degree of product recovery, e.g. product remanufacturing. [...] Another mean of closing the material flow is to focus on Functional sales instead of selling physical products". Moreover, "this research does not include any environmental calculation of whether the remanufacturing of products is environmentally benign or not [...] hence, this paper focuses on technical and economic aspects of remanufacturing". Therefore, even if this paper starts from the same considerations of Tomiyama (1997) and Umeda et al. (2000), it considers the environmental problem from a more objectivist perspective of ontology. Elements of objectivist (but, to some extents, also subjectivist) epistemology can be found in the following extract: "The phenomenon of functional sales has become more prevalent in current consumer patterns and its emergence is mainly market-driven. In functional sales, a very strong focus in placed on how to fulfil customer needs and create customer value. [...] Within functional sales, the function-providing company decides how to fulfil the function that the customer is buying. [...] In the cases of renting, leasing, and functional sale, the product is not sold and a contract is written between user and provider".

Functional Product

The contribution by Alonso-Rasgado et al. (2004), dealing with functional (total care) product, presents a strong objectivist tendency: it starts with definitions and problem statements getting directly inside the topic with a clear "practical intent". This work is focused on "service design in the context of functional products and makes proposals illustrated by examples". The authors highlight the long-term nature of functional products together with the need for the provider "to be involved in an intimate business relationship with the customer" so as to better design the whole system of hardware and services, and best meet customer requirements (which is both objectivist and subjectivist epistemological positions). Furthermore, more elements proving the objectivist tendency on the ontological side can be retrieved in the following: "There are certain key advantages to functional products. For the customer, they provide continuously competitive products [...] with minimal capital expenditure and guaranteed availability. For the supplier, they provide the opportunity to develop increased intellectual knowledge, generate a 'smoothed' cash flow and provide long-term business stability". The extracted parts reported above demonstrate a clear practical, or to better say, Pragmatic orientation of the work analysed. Indeed, because of its ontological objectivist perception of business, and because of an epistemological view combining subjectivist tendencies with objectivist ones, this work can be classified as a Pragmatist paper.

Just like the previous work by Alonso-Rasgado et al. (2004), Lindström et al. (2012) present a clear practical view of functional products and issues related to their planning, development and management. From the introduction of this work, it clearly emerges an objectivist ontological perspective. The paper focuses on the development process for a functional product; more precisely, it proposes a conceptual development process for functional product, and the research was based on "semi-structured and open-ended interviews." Moreover, "In a functional product development scenario, a function is realised by developing hardware, software, a service support system, and management of operation. [...] A large number of decisions are required, and the management of such a development project can be complex. Given that complexity, developing a functional product likely requires keeping the development of all components needed well integrated and coordinated". Considering the extracted parts, with several elements linked to ontological objectivism and to both epistemological objectivism and subjectivism, the paper can be acknowledged as part of the critical theory stream.

Extended Product

Thoben et al. (2001), who introduced the term extended product, start with a reflection about changes occurred in market dynamics and competition because "advanced logistics and transportation systems have brought different national and international markets closer together. [...] Technical strategies such as e-business, e-commerce, m-commerce or financial strategies such as shareholder-value dominate the discussion of the CEO's work, and put an enormous pressure on enterprises to be agile, flexible and innovative". The authors individuate the opportunity to foster differentiation in what they call "product extension": the differentiation of products is "based on tangible aspects of a product and the intangible extensions based on integrated services." In this way, through the provision of a utility package, called extended product, customer's need can be better met and differentiation and competition can be focused not only on physical characteristics of products.

Definition of extended products must take into account two basic concepts lying behind the consumer's choice: demand and requirements. "Requirements define what are customer's needs whereas demand describes a precise item that will satisfy the requirements". The work is focused towards providing a definition and clear picture of the extended product concept. This is done by studying differences between traditional and extended products under different aspects: traditional concept of products and its extension with extended product, life cycle phases of a traditional product and its extension in the case of extended product, etc. Finally, the paper proposes a layered model in order to describe and represent extended product. Moving from ontological-subjectivist considerations, this paper deeply examines the concept of extended products, in order to depict a clear vision of its characteristics: to do so, it analyses and compares characteristics of traditional products and extended product under different aspects, raising both subjectivist and objectivist epistemological issues. For this reason, the paper can be classified as an interpretivist paper.

Integrated Solution

Davies (2004) talked about integrated solution. In the introduction, the author states that competitive advantage is not simply about providing services, but how services are combined with products to provide high-value integrated solutions that address a customer's business or operational needs. These new trends toward high-value services provision are encouraging firms in renovating their business models, and this will also bring to a renewal in how to reach a source of competitive advantage: "firms are increasingly competing by building on their 'core manufacturing capabilities' and integrating forwards into the provision of high-value services that address each customer's needs". For what concerns the methodology employed to investigate these topics, this contribution draws upon case study research undertaken during a 3-year collaborative research project with five international firms. Davies (2004) discusses the important differences emerged by comparing integrated solutions provided in capital goods compared with consumer goods. Because of its double ontological position, with a clear tendency to objectivist epistemology, the definition by Davies (2004) can be classified, with some reserves, as a half way between Conventionalism and Pragmatism.

Some papers dealing with the same term, like PSS, show the same epistemological view but an opposed ontological view, while those about Post Mass Production Paradigm share the same ontological and epistemological subjectivist perspectives, but with different degrees, like papers dealing with functional product, which are ontologically objectivist and epistemologically subjectivist.

It is possible to see how the terms servitization and servicification, although are quite similar, have been introduced with 20 years of difference (1988 and 2010) and have completely opposite ontological and epistemological views.

The two basic papers regarding PSS (Goedkoop et al. 1999; Mont 2002) have the same epistemological perspective, but different ontological tendencies; this can be explained by the fact that the work by Goedkoop et al. (1999) was a report made by practitioners (working for PWC), commissioned by the Dutch Ministries of Economic Affairs and Environment, while the work by Mont was an academic publication; this fact could explain the different views underlying the research work and could bring to consider the report by Goedkoop et al. (1999) as an outlier, for what concerns the ontological aspect.

The differences between the papers dealing with Post Mass Production Paradigm (Tomiyama 1997; Umeda et al. 2000) are probably due to the fact that the work by Tomiyama (1997) was a precursor of the topic posing itself on a stronger position if compared to the research of the period, while the following work by Umeda et al. (2000) can be located into a more mature research.

Works dealing with functional product show some minor differences only on an epistemological perspective, probably due to different views of the authors.

Many papers are located within, or near, the Pragmatist stream of research: the majority of them show the same ontological and epistemological perspectives, respectively objectivist and subjectivist.

Minor differences within this group can be attributable to different research fields from which the authors come from (e.g. Engineering and Design; Marketing; ICT; Business, Management and Accounting, etc.).

So, the great majority of papers dealing with the topic of servitization (as sometimes is indicated the overall field of research) share an objectivist ontological view, as can be expected by a great number of works dealing with economics and business, but a subjectivist epistemological view. The motivation can be probably found because the PSS has a great focus on users' needs and preferences and on functions' fulfilment and these characteristics imply to reconsider the whole business models implemented under a more subjectivist light.

1.2 The Challenging Transition of Servitization: Integrating and Bundling Products and Services

Companies pursuing a servitization strategy should be aware about all opportunities and challenges deriving from the integration of products and services.

A business strategy based on a PSS establishes a value proposition focused on final users' needs rather than on the product (Baines et al. 2007) allowing for an

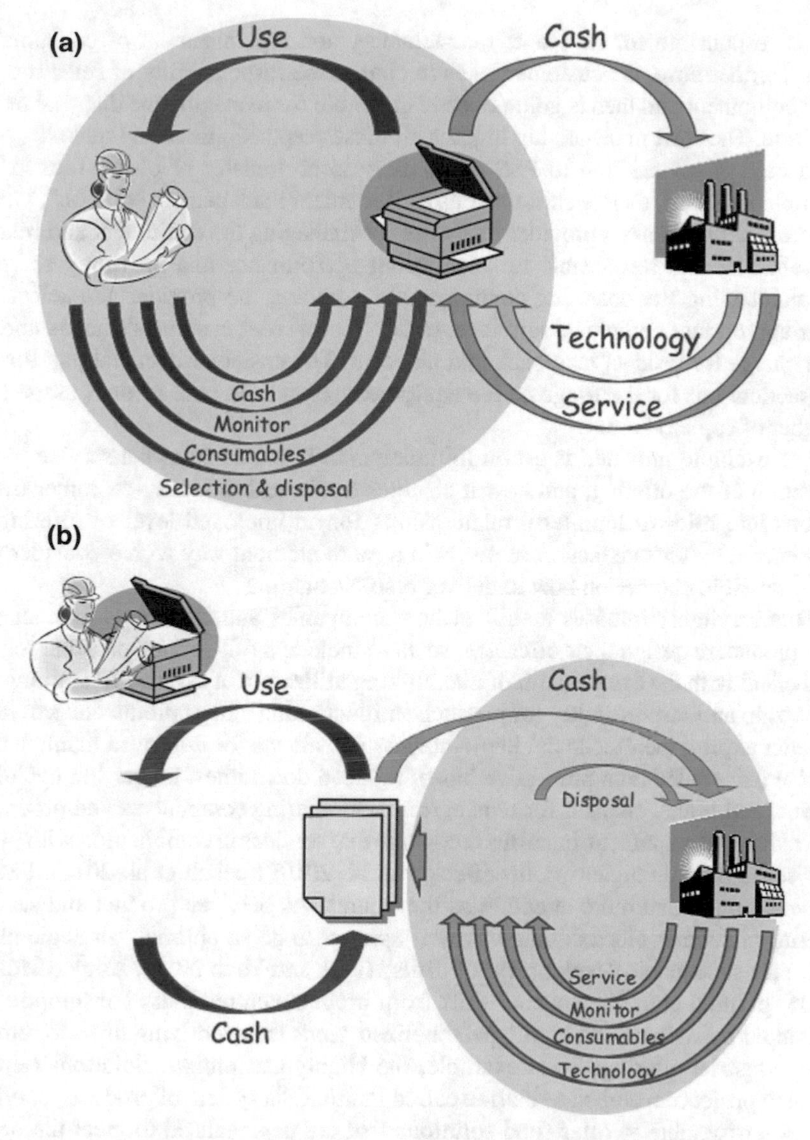

Fig. 1.4 The evolution of value and changes in focus (Lee and AbuAli 2011)

easier design of a need-fulfilment system with radically lower impacts in terms of environmental and social benefits (Mont 2002).

We should consider, for instance, the example reported in Fig. 1.4, where is shown the "servitized transformation" of the traditional purchase of a photocopier. Usually, the producer sells the photocopying machine and a basic service component to ensure its installation and functioning; the customer pays a price and then, after the product is sold and transaction is over, the customer becomes the owner of the photocopier, and is responsible for its usage, maintenance, and replenishment of consumable parts. Furthermore, the customer takes in charge the responsibility of selecting the right equipment and then is going to be responsible for managing the disposal of the machine. The shift in ownership implies all these responsibilities and others.

In case of a transition to PSS offer, there is no transfer of ownership. In the example provided, there a shift towards a "document management solution", where the producer becomes a provider in charge for managing the equipment and related consumables and responsible for monitoring performance and providing services for maintaining the operating conditions. In addition, the provider can select the most appropriate equipment and level of service to meet customer's needs and he is in charge for product take-back and disposal. The customer does not pay for the transaction, but for the usage of the equipment, on a time base or on a usage (i.e. number of copies) base.

The example provided is useful to understand how PSS allows a service-based transition of the offering, and how it changes traditional producer–customer transactions into mid- or long-term relationships for an improved level of offering to customer, a better satisfaction of needs in a more efficient way with a considerable set of possible choices on how to deliver results/solutions.

This inevitably imposes a shift in how companies and customers interact and how producers design their offerings, so as to include a full-service package for the final client with the extra benefit of maximizing utilization of assets. Maintaining the ownership and responsibility for production functioning allows producers/providers to better exploit their technical know-how, which allows for improved maintenance service (scheduled on a preventive base), reduced downtimes, longer life cycles of product and higher chances for reusing/remanufacturing components and products.

Traditional manufacturing firms recognize that services in combination with products could provide higher profits (Becker et al. 2010; Lockett et al. 2011). PSS is attracting more and more attention as the boundaries between product and service offerings becomes blurred: that is why it appears to be an optimal "strategic alternative for sustainable development of firms" (Park and Yoon 2015). As also Morelli (2006) pointed out, "the epochal shift from product-centred mass consumption to individual behaviours and highly personalized needs is now driving firms to rethink their industrial offerings". For example, the Highly Customized Solutions (HiCS) research project developed a solution called Punto X: "a system of products, services and expertise, able to offer food solutions that are personalised to meet the needs of specific contexts-of-use. The personalisation is obtained thanks to the flexibility in the meal composition, the organisation of distribution and delivery systems, and through service/consumer interfaces" (Krucken and Meroni 2006).

PSS allows modern organizations to meet these new evolved needs by also maintaining a clear focus on sustainability needs, which are always more pressing in organizations' core businesses (Cook et al. 2006). In this way, companies can operate a shift in the offerings, securing competitiveness and sustainability at the same time (Azarenko et al. 2009; Beuren et al. 2013).

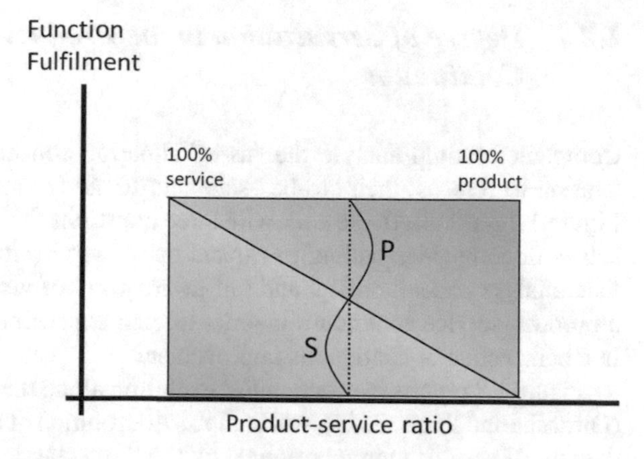

Fig. 1.5 Product–service ratio for a given function/need (Goedkoop et al. 1999)

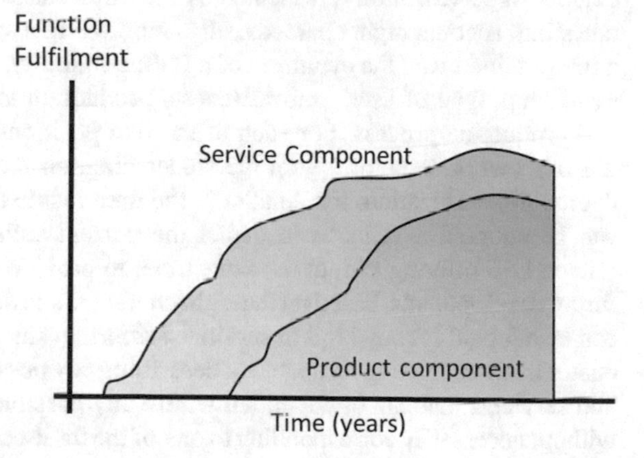

Fig. 1.6 Product–service ratio with time variations (Goedkoop et al. 1999)

As reported in Fig. 1.5, for a given level of customer's satisfaction, there are various possible combinations of product and service components and this can be defined as the product–service ratio, a key characteristic for a PSS.

Furthermore, we can consider the situation depicted in Fig. 1.6 as related to a single moment in the possible evolution over time of a servitized offering, but we must also take into account the evolution of product–service ratio over time, which might also bring variations in the level of customer's satisfaction, as depicted in Fig. 1.6.

The same function (need) can be fulfilled (satisfied) even by different combinations of products and services: this is an example of the potential carried by PSS in pursuing different goals at the same time like, for instance, decoupling the environmental needs from economic performances; for instance, two different PSS offerings might address the same need but, in one case, the presence of a major service component can bring a reduction in material consumption, relevant reduction in material use, production costs and waste production.

1.2.1 *Degree of Servitization in the Product–Service Continuum*

Companies should analyse the “as-is” situation concerning their degree of servitization to forecast their “to-be” state and to study future paths for improvement. Figure 1.7 reports a framework with three questions that managers should ask themselves in companies willing to expand their “servitized base”. A company can, in fact, analyse under a critical and self-aware point of view its current position along a product–service continuum in order to plan expansion paths towards a fixed goal in a perspective of continuous improvement.

Figure 1.8 reports the companies’ evolution along the product–service continuum (Dimache and Roche 2013), where the ideal evolution of a company towards different degrees of servitization represented by the three classic PSS categories. The model takes into account eight characteristics, reported in a spider diagram, to describe in a more refined way the evolution of a PSS: tangibility, product complexity, product ownership, type of user, innovativeness, product durability, customer involvement and production process. For each of the five positions (from A to E) identified in the framework, examples with the spider diagram are provided; the bigger is the degree of servitization, the smaller is the area inside the graph. These dimensions can be adopted as a mean to depict the current situation of a company and the related PSS offering and, at the same time, to provide a more punctual way to plan future developments, based on the eight characteristics. Furthermore, this framework can be adopted for any kind of servitized offering, since it has a very high degree of customization, making it capable of describing any possible combination of products and services, that can be identified within any possible point inside the continuum without necessarily corresponding to one of the three categories (which could always be used as a reference point).

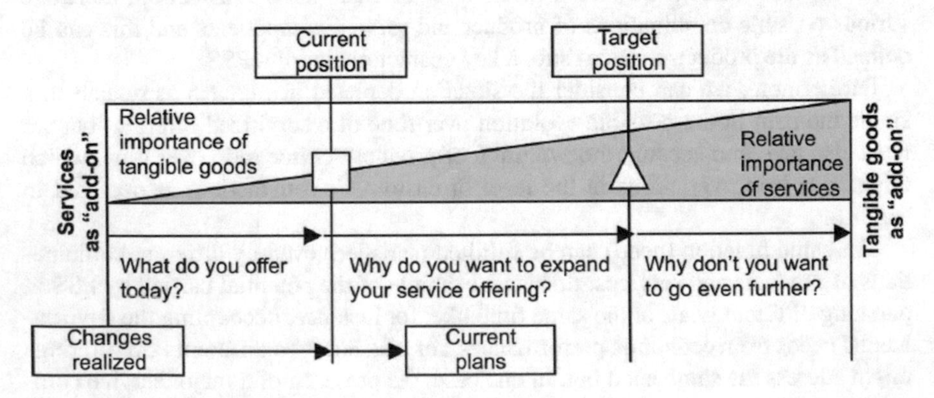

Fig. 1.7 The product–service continuum (Oliva and Kallenberg 2003)

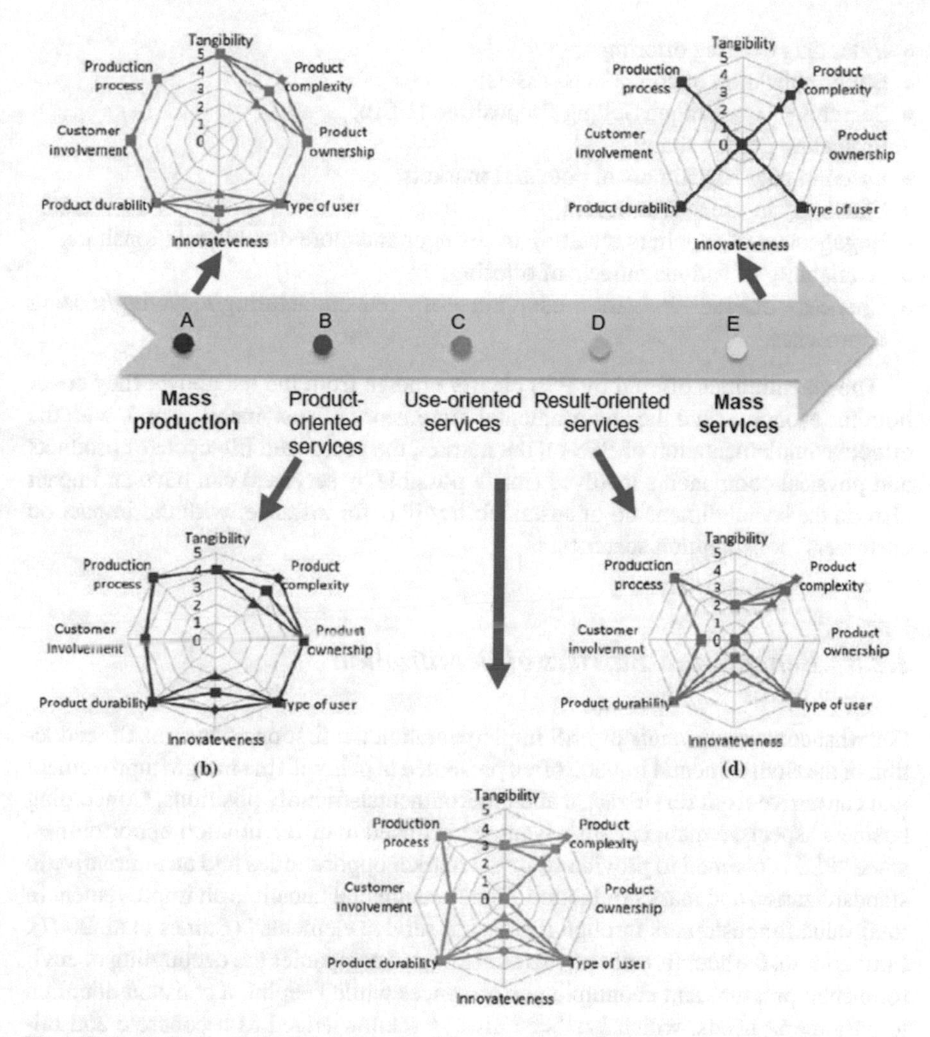

Fig. 1.8 Evolution towards the product–service continuum (Dimache and Roche 2013)

1.2.2 Drivers of Servitization

PSS brings within itself a non-negligible value deriving from various product–service combinations, carrying several different drivers for companies. These drivers might be valuable for every kind of company (product manufacturers and service companies) going through servitization adoption for:

- Building strong and durable relationships with customers;
- Cooperating with authorities to achieve advances in legislation and foster adoption of environmental-friendly solutions;
- Reducing environmental costs, first of all linked to waste production;

- Extending existing offerings;
- Better utilization of companies' assets;
- Searching for a Unique Selling Proposition (USP);
- Protecting market share;
- Discouraging newcomers in potential markets;
- Flexibility in use and/or in rent;
- Engagement of suppliers resulting in stronger and more durable relationships;
- Availability of various models of offering;
- Chances offered by the adoption of remanufacturing/recycling/reusing approaches.

The potentialities offered by PSS clearly emerge from the list above: they cover both the economic and the environmental dimensions of sustainability and, with the effective implementation of PSS on the market, the prolonged life cycle of products and physical components involved (made possible by services) can have an impact also on the social dimension of sustainability like, for instance, with the impact on customers' consumption schemes.

1.2.3 Benefits and Barriers of Servitization

For what concerns *benefits* of PSS implementation the first one concerns the reduction of the environmental impact, often presented in pair with the image improvement that can derive from servitization and environmental-friendly positions. Concerning business aspects, a main benefit/advantage is linked to differentiation opportunities, since "PSS is claimed to provide strategic market opportunities and an alternative to standardization and mass production. The fundamental benefit is an improvement in total value for customers through increasing service elements" (Baines et al. 2007). Furthermore, the adoption of servitization allowed companies the decoupling of environmental pressure and economic performances while keeping a constant attention to customers' needs, which has been always acknowledged as a concrete and relevant strategic opportunity. More in detail, different authors (Tukker and Tischner 2006; Baines et al. 2007; Sundin et al. 2009; Aurich et al. 2009; Mittermeyer et al. 2011; Tan 2010) separately considered benefits delivered to customers and benefits delivered to companies.

For consumers:

- Higher value delivered.
- The degree of service flexibility;
- The degree of personalisation offered;
- Higher quality level;
- Improved satisfaction of needs;
- Offering of new functionalities, thanks to combinations products and services;
- No concerns linked to monitoring product status;
- No concerns for end-of-life disposal;

For companies:

- Creation of new market opportunities;
- Disclosure of new sources of competitive advantage;
- Availability of detailed information on the usage of products and their performance;
- Higher margins provided by service replacement of products;
- Stronger relationships with customers bringing to a higher level of customer retention and trust;
- Disclosure of new innovation potential, thanks to the service elements in the offering;
- Chances for the reuse/remanufacture of products and components.

Furthermore, PSS can bring benefits that directly impact the environmental and social dimensions of sustainability (Baines et al. 2007) like, for instance:

- Reduction in consumption of inputs;
- Reduction in the production of wastes and by-products;
- Public pressure on environmental issues;
- Increase in service supply;
- Chance for new job creation and development.

There are also a non-negligible number of *barriers* to PSS adoption as well. Main resistances to servitization shift mainly come from customers and companies' employees. Customers exhibit resistances in changing their consumption habits and consumption schemes, mainly for what concerns the most radical examples of PSS like, for instance, offerings of use-oriented and/or result-oriented categories (Ceschin and Vezzoli 2010). These changes to consumption schemes, besides, do not always bring significant benefits as expected, posing a new element of risk in the overall process of servitization. The reasons for this resistance to the development of more advanced services are many, and the academic literature has tried to identify them over the years. Some barriers can be found also in the new role of customer.

The introduction of a PSS imposes also a shift in organizational cultures, mainly related to a change in the conception of business value. Furthermore, the adoption of PSS-related offerings brings quite often the need for new pricing policies and a non-negligible risk linked to these policies. There are also risks concerning the lack of experience in service design and service delivery for many manufacturing companies, as well as the lack of technological know-how for service companies, which in pair bring to the need for skilled personnel (Cavalieri and Pezzotta 2012).

Generally, the two main barriers presented often occur together with the resistance in acceptance from stakeholders, especially for partners and suppliers operating in the supply chain of the servitized firm. Cooperation of these actors is a key element in ensuring a successful PSS adoption and development, so as to ensure a win-win-win (supplier–producer–customer) strategy (Annarelli et al. 2016). This commitment is essential because of the changes required in the supplier–producer relationship, passing from a transactional relationship to a long-term one.

1.2.4 The Service Paradox

Another crucial element that acts as a barrier to the development of advanced services is the widespread fear among companies of incurring in the so-called "service paradox" (Gebauer et al. 2005): it is well established that increasing servitization leads to an increase of revenues, but it does not always coincide with an increase in profits; as observed in numerous cases, the provision of services often implies an increase in fixed costs, which, together with the poor scalability of servitization, can go to erode most of the profits making the adoption of this business model counterproductive.

The advent of digitalization has, however, contributed to relax this barrier, making the transition to a PSS policy more scalable and less traumatic. Thanks to technological advancement, companies are now able to opt for a gradual "servitization" of their value chain (Coreynen et al. 2018) being able to evaluate whether to go in this perspective to evolve (1) the back-office area (using digital tools to optimize the production efficiency of its organization, and using the acquired knowledge to offer consultancy services to customers) or (2) the front office (focusing instead on the development of digital interfaces that involve customers in the development of the offer and, at the same time, provide them with tools for viewing and managing their data). In general, however, to develop advanced services, and have profits, companies need to renew their delivery system in depth, making it capable of managing the new costs and risks that a PSS offer implies, which requires significant resources.

There are different macro-areas of intervention in this regard, which can change the reality of the organization: from relationships with suppliers, through development of human resources skills, up to a re-engineering of the organization's processes. It is particularly important to have a network of facilities located close to customers, to provide services efficiently and establish a climate of cooperation with the customer (Baines and Lightfoot 2013).

Do not rely on external service providers and manage "in house" the interface with the customer is, in fact, highly recommended for servitized business models, as through it, you can capture valuable data on him and establish a climate of cooperation (Kowalkowski and Brehmer 2008).

This requirement, which needs a large investment in assets and represents an important economic barrier for companies wishing to approach the model of servitization, can now be mitigated by the new technologies introduced with digitalization, which through remote monitoring and control systems they no longer require such a widespread presence on the territory, enabling the possibility of providing services even from a distance.

1.2.5 *The New Role of Client*

The complex nature of servitization and PSS depends mainly from the plurality of elements concurring in their definition, in their design/development, and in all necessary steps to address customers' needs.

The analysis of customers' needs is the starting and arrival point of the whole servitization process companies cannot ignore the role played by final clients, since most of times they are actively involved in the delivery of the servitized solution/PSS. Indeed, in Chap. 2 the central role that customers play in the servitization context will be discussed, being not simply anymore agent to which deliver the final product, but part of the value creation process. Servitization and PSS are closely related to the concept of *value co-creation*: customers have an active role in the service delivery process, since the simultaneous production and consumption of service due to its intangible nature. In several business models focused on sharing economy concepts, the participation of customers is a key point. That is why, in the context of servitization and PSS, there has been also a rise in attention to contracts. Furthermore, resistance to change and/or acceptance from customers, is another key element in determining the success of a PSS.

Two different types of customers can be identified (Carlborg et al. 2018): passive and active. Passive customers mostly rely on provider's capabilities for service delivery, since they lack time, money or incentives to actively participate and be involved in service deployment. In these cases, there is a low level of direct interaction between customer and service provider with technology playing a relevant mediating role. On the other hand, there are active customers who, led by stronger drivers, have a direct participation in the service delivery process. This is mostly the case of tailored solutions with a high level of customization, which starts from the design phase. Furthermore, this is likely to happen more frequently in B2B contexts rather than in B2C.

The types of customers determine the quality of services that can be offered (Baines and Lightfoot 2013). Based on the value proposition that organizations develop with their customers, there are:

- Customers who want to do it themselves: they have no intention of undertaking a cooperative path with their product supplier and, therefore, rely on it only for basic services, such as the supply of the product and spare parts.
- Customers who want to do it with them: in addition to supplying the product, they also rely on the supplier to request intermediate level services, such as significant repairs and revisions. In this case, the relationship between the two parties does not end with the sale and shipment process of the goods, but it also continues during the post-sale phase, albeit in a very superficial manner.
- Customers who want us to do it for them: they contract with the supplier only the capacity and the performances that must be supplied, and letting it to take the load of a large part (if not all in some cases) of the asset management activities. In this case, we talk about advanced services, which are also those that have the greatest potential and benefits for both parties.

Therefore, one of the most important limits to the adoption of a pushed servitization is often just the "acceptance of the new model by customers": it is not easy to convince a customer, used to buy simply a physical asset, that is convenient for him to pay an extra fee to get complementary services, or pay to simply get a performance (Baines et al. 2007). This limit certainly represents, together with the resistance to change within the company, the greatest barrier to the development of a business model based on servitization.

Over time, however, the market has evolved, and we are witnessing a growing change in the mentality of customers, who increasingly show that they have evolved their concept of consumption: from coinciding with the purchase of a physical product, to the acquisition of a performance (Gao et al. 2011).

Seven Key Facts

- Servitization indicates a shift towards the development of product–service mixed offering with the aim of replacing product selling.
- Product service system (PSS) is constituted by a plurality of elements and characteristics resulting in a considerable variety of options and different degrees of servitization.
- In product-oriented PSS, the focus is still on product selling with extra services added to the offering.
- In use-oriented PSS, the customer pays (usually according to a time unit) for using the product with no shift in ownership.
- In result-oriented PSS, the customer pays for the delivery of a functional result.
- Servitization might be characterized by "paradoxes" that can undermine the realization of profits.
- One of the most important limits in the adoption of a pushed servitization is often just the "acceptance of the new model by customers".

References

T. Alonso-Rasgado, G. Thompson, B.O. Elfström, The design of functional (total care) products. J. Eng. Des. **15**(6), 515–540 (2004)

A. Annarelli, C. Battistella, F. Nonino, Product service system: a conceptual framework from a systematic review. J. Cleaner Prod. **139**, 1011–1032 (2016)

J.C. Aurich, N. Wolf, M. Siener, E. Schweitzer, Configuration of product-service systems. J. Manuf. Technol. Manage. **20**(5), 591–605 (2009)

A. Azarenko, R. Roy, E. Shehab, A. Tiwari, Technical product-service systems: some implications for the machine tool industry. J. Manuf. Technol. Manage. **20**(5), 700–722 (2009)

T.S. Baines, H. Lightfoot, E. Steve, A. Neely, R. Greenough, J. Peppard, R. Roy, E. Shehab, A. Braganza, A. Tiwari, J. Alcock, J. Angus, M. Bastl, A. Cousens, P. Irving, M. Johnson, J. Kingston,

H. Lockett, V. Martinez, P. Michele, D. Tranfield, I. Walton, H. Wilson, State-of-the-art in product-service systems. Proc. Inst. Mech. Eng. Part B: J. Eng. Manuf. **221**(1), 1543–1552 (2007)

T. Baines, H.W. Lightfoot, Servitization of the manufacturing firm. Int. J. Oper. Prod. Manage. **34**(1), 2–35 (2013)

G. Baxter, I. Sommerville, Socio-technical systems: from design methods to systems engineering. Interact. Comput. **23**(1), 4–17 (2011)

J. Becker, D.F. Beverungen, R. Knackstedt, The challenge of conceptual modeling for product-service systems: status-quo and perspectives for reference models and modeling languages. IseB **8**, 33–66 (2010)

F.H. Beuren, M.G.G. Ferreira, P.A.C. Miguel, Product-service systems: a literature review on integrated products and services. J. Cleaner Prod. **47**, 222–231 (2013)

M. Boehm, O. Thomas, Looking beyond the rim of one's teacup: a multidisciplinary literature review of product-service systems in information systems, business management, and engineering & design. J. Cleaner Prod. **51**, 245–260 (2013)

T. Briceno, S. Stagl, The role of social processes for sustainable consumption. J. Cleaner Prod. **14**, 1541–1551 (2006)

P. Carlborg, D. Kindstrom, C. Kowalkowski, Servitization practices: a co-creation taxonomy, in *Practices and Tools for Servitization—Managing Service Transition*, ed. by Kohtamaki et al. (Palgrave MacMillan, Cham, 2018), pp. 309–321

S. Cavalieri, G. Pezzotta, Product-Service Systems Engineering: State of the art and research challenges. Comput. Ind. **63**, 278–288 (2012)

F. Ceschin, C. Vezzoli, The role of public policy in stimulating radical environmental impact reduction in the automotive sector: the need to focus on product-service system innovation. Int. J. Autom. Technol. Manage. **10**(2–3), 321–341 (2010)

F. Ceschin, How the design of socio-technical experiments can enable radical changes for sustainability. Int. J. Des. **8**(3), 1–21 (2014)

M. Cook, T.A. Bhamra, M. Lemon, The transfer and application of product service systems: from academia to UK manufacturing firms. J. Cleaner Prod. **14**, 1455–1465 (2006)

W. Coreynen, P. Matthyssens, R. De Rijck, I. Dewit, Internal levers for servitization: how product-oriented manufacturers can upscale product-service systems. Int. J. Prod. Res. **56**(6), 2184–2198 (2018)

A. Davies, Moving base into high-value integrated solutions: a value stream approach. Ind. Corp. Change **13**(5), 727–756 (2004)

E. Dewberry, M. Cook, A. Angus, A. Gottberg, P. Longhurst, Critical reflections on designing product service systems. Des. J. **16**(4), 408–430 (2013)

A. Dimache, T. Roche, A decision methodology to support servitization of manufacturing. Int. J. Oper. Prod. Manage. **33**(11–12), 1435–1457 (2013)

European Commission, *For a European Industrial Renaissance, Communication from the Commission to the European Parliament, The Council, The European Economic and Social Committee and the Committee of the Regions.* (COM/2014/014 final). Brussels (2014)

S. Evans, P.J. Partidário, J. Lamberts, Industrialization as a key element of sustainable product-service solutions. Int. J. Prod. Res. **45**(18–19), 4225–4246 (2007)

J. Firnkorn, M. Müller, What will be the environmental effects of new free-floating car sharing systems? The case of car2go in Ulm. Ecol. Econ. **70**(8), 1519–1528 (2011)

C.A. Friebe, P. von Flotow, F.A. Täube, Exploring the link between products and services in low-income markets—evidence from solar home systems. Energy Policy **52**, 760–769 (2013)

J. Gao, Y. Yao, V.C.Y. Zhu, L. Sun, L. Lin, Service-oriented manufacturing: a new product pattern and manufacturing paradigm. J. Intell. Manuf. **22**, 435–446 (2011)

H. Gebauer, E. Fleisch, T. Friedli, Overcoming the service paradox in manufacturing companies. Eur. Manage. J. **23**(1), 14–26 (2005)

M.J. Goedkoop, C.J.G. van Halen, H.R.M. te Riele, P.J.M. Rommens, *Product service systems, ecological and economic basics*. Report for the Dutch ministries of Economic Affairs and of Environment (1999)

M. Halme, C. Jasch, M. Scharp, Sustainable homeservices? Toward household services that enhance ecological, social and economic sustainability. Ecol. Econ. **51**(1–2), 125–138 (2004)
H. Komoto, N. Mishima, T. Tomiyama, An integrated computational support for design of system architecture and service. CIRP Ann. Manuf. Technol. **61**, 159–162 (2012)
C. Kowalkowski, P.O. Brehmer, Technology as a driver for changing customer-provider interfaces. Manage. Res. News **31**(10), 746–757 (2008)
M.E. Kreye, L.B. Newnes, Y.M. Goh, Uncertainty in competitive bidding—a framework for product–service systems. Prod. Plann. Control **25**(6), 462–477 (2014)
L. Krucken, A. Meroni, Building stakeholder networks to develop and deliver product-service systems: practical experiences on elaborating pro-active materials for communication. J. Cleaner Prod. **14**, 1502–1508 (2006)
G. Lay, M. Schroeter, S. Biege, Service-based business concepts: a typology for business-to-business markets. Eur. Manage. J. **27**(6), 442–455 (2009)
J. Lee, M. AbuAli, Innovative Product Advanced Service Systems (I-PASS): methodology, tools, and applications for dominant service design. Int. J. Adv. Manuf. Technol. **52**, 1161-1173 (2011)
J. Lindström, M. Löfstrand, M. Karlberg, L. Karlsson, A development process for functional products: hardware, software, service support system and management of operation. Int. J. Prod. Dev. **16**(3–4), 284–303 (2012)
H. Lockett, M. Johnson, S. Evans, M. Bastl, Product service systems and supply network relationships: an exploratory case study. J. Manuf. Technol. Manage. **22**(3), 293–313 (2011)
M. Lodefalk, *Servicification of manufacturing—evidence from Swedish firm and enterprise group level data*. Working Paper, Swedish Business School at Örebro University (2010)
E. Manzini, C. Vezzoli, A strategic design approach to develop sustainable product service systems: example taken from the 'environmentally friendly innovation' Italian prize. J. Cleaner Prod. **11**, 851–857 (2003)
D. Maxwell, R. van der Vorst, Developing sustainable products and services. J. Cleaner Prod. **11**, 883–895 (2003)
H. Meier, R. Roy, G. Seliger, Industrial product-service system—IPS2. CIRP Ann. Manuf. Technol. **59**, 607–627 (2010)
S.A. Mittermeyer, J.A. Njuguna, J.R. Alcock, Product–service systems in health care: case study of a drug–device combination. Int. J. Adv. Manuf. Technol. **52**, 1209–1221 (2011)
O. Mont, Clarifying the concept of product-service system. J. Cleaner Prod. **10**, 237–245 (2002)
N. Morelli, Developing new product service systems (PSS): methodologies and operational tools. J. Cleaner Prod. **14**, 1495–1501 (2006)
A. Neely, Exploring the financial consequences of the servitization of manufacturing. Oper. Manage. Res. **1**, 103–118 (2009)
N. Nishino, S. Wang, N. Tsuji, K. Kageyama, K. Ueda, Categorization and mechanism of platform-type product-service systems in manufacturing. CIRP Ann. Manuf. Technol. **61**, 391–394 (2012)
R. Oliva, R. Kallenberg, Managing the transition from products to services. Int. J. Serv. Ind. Manage. **14**(2), 160–172 (2003)
H. Park, J. Yoon, A chance discovery-based approach for new product-service system (PSS) concepts. Serv. Bus. **9**, 115–135 (2015)
A. Richter, T. Sadek, M. Steven, Flexibility in industrial product-service systems and use-oriented business models. CIRP J. Manuf. Sci. Technol. **3**, 128–134 (2010)
R. Rivas-Hermann, J. Köhler, A.E. Scheepens, Innovation in product and services in the shipping retrofit industry: a case study of ballast water treatment systems. J. Cleaner Prod. **106**, 443–454 (2015)
R. Roy, Sustainable product-service systems. Futures **32**, 289–299 (2000)
E. Sundin, B. Bras, Making functional sales environmentally and economically beneficial through product remanufacturing. J. Cleaner Prod. **13**, 913–925 (2005)
E. Sundin, M. Lindahl, W. Ijomah, Product design for product/service systems—design experiences from Swedish industry. J. Manuf. Technol. Manage. **20**(5), 723–753 (2009)

A.R. Tan, *Service-oriented product development strategies: Product/Service-Systems (PSS) development*. DTU Management Ph.D. Thesis. Kgs. Lyngby. (2010)

T. Tasaki, S. Hashimoto, Y. Moriguchi, A quantitative method to evaluate the level of material use in lease/reuse systems of electrical and electronic equipment. J. Cleaner Prod. **14**, 1519–1528 (2006)

K.D. Thoben, J. Eschenbächer, H. Jagdev, Extended products: evolving traditional product concepts, in *7th International Conference on Concurrent Enterprising*, vol. 7 (2001), pp. 429–439

T. Tomiyama, A manufacturing paradigm toward the 21st century. Integr. Computer-Aided Eng. **4**(3), 159–178 (1997)

A. Tukker, Eight types of product-service system: eight ways to sustainability? Experience from SusProNet. Bus. Strategy Environ. **13**, 246–260 (2004)

A. Tukker, U. Tischner, Product-service as a research field: past, present and future. Reflection from a decade of research. J. Cleaner Prod. **14**, 1552–1556 (2006)

Y. Umeda, A. Nonomura, T. Tomiyama, Study on life-cycle design for the post mass production paradigm. Artif. Intell. Eng. Des. Anal. Manuf. **14**, 149–161 (2000)

S. Vandermerwe, J. Rada, Servitization of business: adding value by adding services. Eur. Manage. J. **6**(4), 314–324 (1988)

Chapter 2
The New Role of Client: From Ownership to Value Co-creation

This chapter focuses on the key role played by customers in product service system context. As evidenced at the end of Chap. 1, customers and their needs are the starting point of a PSS-based proposal, and this is also a focal point for the value proposition at the core of servitization and related business models. Furthermore, in PSS, there is no longer the traditional process of value creation and delivery to clients, but there is an "all-around" involvement of customers through value co-creation, going from participation in design phase to the delivery phase of the product–service offering. Key element in this context is also the changing concept of ownership, since many PSS offerings do not imply a shift in ownership (like in traditional product selling) with customers paying directly for the usage and/or performance connected to physical products.

In this chapter, the new role of customers will be investigated first of all under the perspective of elements to be taken into account when designing/developing a PSS and, then, under the perspective of how customers should be managed when implementing and delivering a PSS.

2.1 Servitization as a New Value Proposition

Figure 2.1 servitization represents a new value proposition key element in this context is also the changing concept of ownership, since many PSS offerings do not imply a shift in ownership (like in traditional product selling) with customers paying directly for the usage and/or performance connected to physical products. The value proposition in the PSS concerns the value that is offered by integrating product and service. Typical examples of value are the reduced responsibility on product durability and the guarantee of functionality (Isaksson et al. 2009). Since the PSS provider is usually responsible for operations such as maintenance and repair, reducing operational costs can be understood as a form of value proposition (Alonso-Rasgado et al. 2004). This type of activity does not increase the tangible and intrinsic value, but increases its intangible value linked, for example, to the values of trust, the commitment to

A. Annarelli et al., *The Road to Servitization*,
https://doi.org/10.1007/978-3-030-12251-5_2

attractiveness (Grönroos 2011). The definition and perception of value depends on the type of stakeholder and on its role within the supply chain, on the way in which the service is administered and on its responsibilities (for example, the difference in the perception of the value of a product depending on whether it is purchased or used in leasing (Fishbein et al. 2000)). The definition of the value proposition therefore goes beyond understanding what the service can offer and how a coherent portfolio is developed (Kindström and Kowalkowski 2014). The implementation of the PSS logics calls into question the whole concept of value; if traditionally it was linked to the exchange phase, it is now linked to the use phase (Vargo and Lusch 2004; Ng et al. 2009; Grönroos 2011). In this sense, value can be the result of different configurations of the value proposition (i.e. Tukker 2004; Smith et al. 2012); for example, the client can positively perceive the possession of the asset, or he can consider advantageous to enjoy its use without facing the costs associated with the purchase (Kujala et al. 2010; Barquet et al. 2013; Reim et al. 2015). So the concept of value in PSS can be declined according to four categories: service offerings, customer value, value co-creation and product ownership. These categories will be described in the following paragraphs.

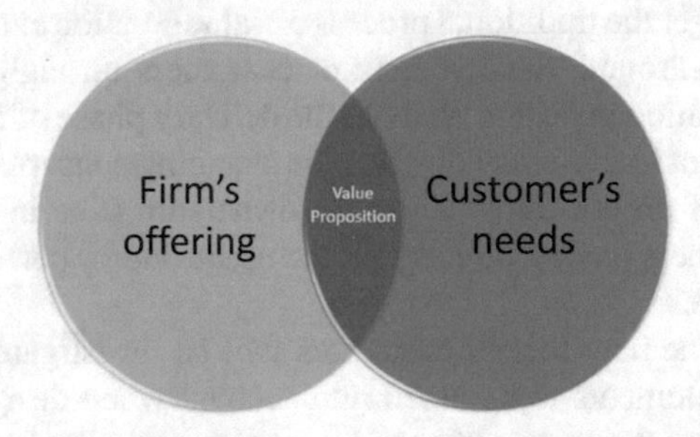

Fig. 2.1 The concept of value proposition

2.1.1 Service Offering

The service sector is an extremely heterogeneous category, and it is possible to find many differences ranging from simple field services to broader services involving more actors (Kindström and Kowalkowski 2014). Given the extent of the potential demand, it is important for the company to develop a portfolio with solutions that are mutually coherent. The extension of the offer with service components is a key factor in the supply of PSS. As a demonstration, we can consider how the evolution of customers' needs and requirements has encouraged providers to develop their skills in offering business and financial services that are particularly useful in the preliminary negotiation phases. These services allow to guide the client on how to plan, design and finance the purchase of a product, its use and maintenance

(Davies 2004). Within this context, customers are differentiated by the importance of their needs in terms of advice during the life cycle of the product. The weaker are the capabilities of the customer, the sooner he will require technical assistance to the provider. Indeed, there can be customers who need to be supported from the initial stages of negotiation up to smart buyers who can rely on much more robust internal capabilities. Financial services also play a key role during the negotiation phase, especially when the customer requests financial assistance for the purchase of extremely expensive products. A well-known practical example is represented by ABB, which offers its clients contracts for sharing value, guaranteeing a decrease in the purchase value and obtaining in exchange for a percentage of the customer's profits (Davies 2004).

A classification of services can distinguish between services that support the functionality of the product provided (e.g. the classic after-sales services), and the services that support the customer's activities related to the product (for example, training courses for a correct use). The first type of service follows the concept traditionally offered on the market, while the second one requires a more advanced and structured product–service perspective (Mathieu 2001). The main purpose of a service provided to support a product is to ensure its functionality and facilitate access for the customer. On the other hand, by offering a service that supports the customer's action, the supplier aims to analyse the ways in which to support particular customer initiatives and to configure in an appropriate manner its organizational structure. This type of classification emphasizes that those responsible for selling advanced services need to have a great knowledge of the customer's production processes and of how the service offered will support its activities.

As can be seen from Table 2.1, these two types can be declined according to four different key points: the recipient who receives the service, the intensity of the relationship, the level of customization and the main elements that characterize it. For a *service to support the product*, the recipient of the service is the product itself, while in the case of an *service supporting customer* it will be a person. The intensity of the relationship is low for a service supporting product and high for a service supporting customer, given: (1) the potential number of people and departments involved, (2) the level of involvement between the parties and (3) the trust that underlies these relationships. Services supporting product are standardized, while services supporting customer are highly personalized. Finally, for services supporting product, the key variables are physical characteristics (tangible components) and processes, while the human variables in the services supporting customer (customer and supplier personnel) have a greater impact. However, this model can be refined by adding a second dimension to classify services based on how the value proposition is established (i.e. distinguishing on the fact that service delivery is guaranteed (input-based) rather than a de-terminated performance (output-based)). Combining the two categories, we obtain four classifications useful for understanding the variety of services offered on the market (Ulaga and Reinartz 2011):

- product life-cycle service,
- process support service,

Table 2.1 Different characteristics of services that support product's functionality (service supporting product) and services that support customer's activities related to the product (service supporting customer)

CHARACTERISTIC	Service Supporting Product	Service Supporting Customer
RECIPIENT	Product	Customer
INTENSITY OF THE RELATIONSHIP	Low	High
CUSTOMIZATION	Low	High
MAIN ELEMENTS	Physical components	Human interaction

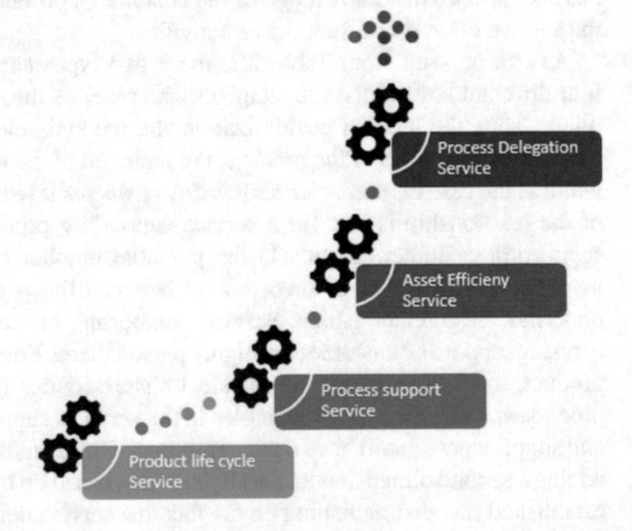

Fig. 2.2 Representation of solutions offered (Ulaga and Reinartz 2011)

- asset efficiency service and
- process delegation service.

These services are shown in Fig. 2.2 and described in Table 2.2.

Table 2.2 Classification of solutions offered (Ulaga and Reinartz 2011)

VALUE PROPOSITION	Focus on the product	Focus on customer's process
INPUT-BASED LOGIC	**Product Life Cycle Service**	**Process Support Service**
DEFINITION	Services provided to facilitate the use of the well and guarantee its functionality	Services that support the customer in managing a process
EXAMPLE	Spare parts supply	Consultancy for logistics management
KEY CAPABILITY	Design to service	Ability to offer Hybrid products
REQUIRED RESOURCES	Service organization	Use of installed base and management of collected data
OUTPUT-BASED LOGIC	**Asset Efficiency Service**	**Process Delegation Service**
DEFINITION	Services designed to guarantee an increase in productivity	Services designed to guarantee a managed process on behalf of the client
EXAMPLE	Remote monitoring	Fleet maintenance dedicated to customer logistics
KEY CAPABILITY	Risk Management	Design to service
REQUIRED RESOURCES	Product development	Sales network

Product Life-cycle Services

Product life-cycle services refer to a wide range of services that facilitate customer access to the good offered by the supplier and ensure its main functionality during all phases of product life before, during and after the sale. This type of service is directly related to the good provided, so the value proposition derives from the classic definition of service: to perform an action on behalf of the client. For example, if a cooling pump in a nuclear plant breaks down, the pump supplier promises to repair it in a very short period of time (Ulaga and Reinartz 2011). This type of product represents a must-have for customers and that there is a lack of propensity on their side to satisfy it. Given the difficulty of differentiating these types of services, the managers have attempted to standardize the product life-cycle services. However, many managers consider this type to be very important as it is possible to build up a reputation as a supplier through good delivery. These characteristics have important implications in the definition of the price of product life-cycle services. Many companies could provide these services for free to secure the sale or to develop a "break it, fix it" logic. To avoid the issue related to the pricing of these services, they are often merged into an "all-inclusive" formula.

Asset Efficiency Services

Given the standardized nature of *product life-cycle services*, it is difficult to differentiate its offer to gain a more competitive position. Many companies have therefore opted to develop new services able to offer added value through the evolution towards *asset efficiency services*, with which we mean the range of solutions aimed at ensuring the productivity of the assets in which the client invested. In the study proposed by Ulaga and Reinartz (2011), the companies specialized in *asset efficiency services* are engaged in activities such as preventive maintenance of ball bearings, field monitoring of moulding presses and customization of robotics software. Similar to what is seen for product life-cycle services, efficient asset services are related to the good provided and are rarely provided as services itself. For example, they cite the case of a manufacturer of medical scanners, who guarantees this type of service only for its own equipment and not for that of its competitors. The comparison between product life-cycle services and efficient asset services reveals several key differences. The transition between *product life-cycle services* and *asset efficiency services* involve a change in the new proposition that moves from the promise of a specific action (the installation of a machine) to the promise of a certain performance (the conformity of 99.8% of the pieces produced). Second, *asset efficiency services* solutions are more customized and allow the provider to differentiate their offer. The third dissimilarity consists in the fact that the *asset efficiency services* are not perceived as fundamental by the client.

Process Support Services

The two previous categories were focused on services connected to the good provided by the supplier, while now the focus is on services aimed at supporting certain processes (*process support services*). This type of activity is oriented towards the customer's production processes, not to the good itself. However, the tendency to provide *process support services* emerges in conjunction with its instruments, although there is a non-negligible share that, in some cases, offers service coverage regardless of the type of instrumentation. In other words, *process support services* are geared to ensure small tasks to support customer processes without taking responsibility for process outputs. The skills related to the management of process services allow providers to emerge and stand out in the market given their strong personalization. If, for example, welding gas is considered a commodity, the knowledge of the supplier on its use during the process can actually be a distinguishing factor. In this case, the propensity to purchase is decidedly high and, usually, the pricing follows the same rules of professional services.

Process Delegation Services

The fourth category in analysis is that of *process delegation services*, defined as that set of services aimed at managing a process on behalf of the client. In this case, the management of the process is totally delegated to the supplier who no longer needs to guarantee only the input, but precise and specific production performances. Given the complexity in handling such solutions, only few companies have entered this context and usually they are leaders in their field.

2.1.2 Customer Value

The concept of value for the client and its related analysis is fundamental for PSS (Payne and Holt 2001; Mont 2002; Vargo and Lusch 2004; Pawar et al. 2009).

The value for the customer is the set of benefits that the company is able to transmit to the customer and can consist in the reduction of the initial investment (the possibility of use formulas that disregard the purchase allows to not immobilize capital), in the minimization of operating costs (due, for example, to maintenance, repairs, upgrades or periods of non-availability of the good caused by breakages) to decreased customer responsibility on the cycle of the product (think, for example, the advantages due to the possibility of leaving the logistics costs of the disposal phase to the provider) (Morris et al. 2005; Isaksson et al. 2009; Barquet et al. 2013).

If those seen above are more tangible advantages, the client may be attracted by other aspects of a different nature that may constitute a substantial share of the value of the good or service. In fact, compared to the traditional products, PSS implies a strong customization that leads to a personalized and unique development of the product, which allows to transmit an added value for the customer who can thus enjoy preferential relations with the provider for reducing efforts to make the purchased services truly operational (Tukker and Tischner 2006). We can identify key elements that can contribute to the creation of the value proposition. They are basically (Fig. 2.3) performance, customization, "getting the job done", cost reduction, risk reduction, usability and flexibility in contracts.

- ***Performance***: for a long time, increasing and guaranteeing the high performance of products was a widespread way to generate value for the customer. A problem regarding the increase in performance may differ between different customer segments and aspects such as price and ease of use become fundamental.
- ***Customization***: by offering integrated solutions of products and services, value can be created to meet specific needs of a single customer or a single segment. The concept of mass customization and co-creation has gained greater importance in recent years. In order to create beneficial interactions both for customers and suppliers, the company must decide whether this is the right path to follow (or not) based on quality and price criteria. The decision to customize products and services for a wide audience and a narrow segment becomes strategic for the impact that economies of scale can have (Osterwalder and Pigneur 2010).
- "***Getting the job done***": by offering solutions that help the client performing a task, the value can be created in various ways. This means that a company offers a product or service to facilitate the work of others. Rolls-Royce is a good example of

Fig. 2.3 The customer value and its key elements

this way of creating value. Its customers rely on Rolls-Royce for the construction and maintenance of engines, allowing the companies to focus only on aspects related to the satisfaction of their customers' needs.

- ***Cost reduction***: the reduction of costs of various kinds allows to attribute an evident added value to the solutions offered. The communication of this opportunity to customers becomes strategic.
- ***Risk reduction***: risk reduction is also perceived positively. To guarantee this, the provider takes on a larger share of responsibility allowing the customer to use equipment without taking on the related risks. An example is the practice of guaranteeing a year of service and maintenance in the automotive sector.
- ***Usability***: for a customer who buys a new solution, it is important the ease of use, so that he can immediately enjoy the benefits related to usage and save time and money related to the training of human resources interfacing process.
- ***Contract flexibility***: when a provider proposes a package of solutions to customers or to a specific market segment, the latter may have different contractual solutions. For example, they can choose whether to take a greater share of risk by buying the asset and exclusively enjoying services related to maintenance and repairs or totally outsourcing the process to guarantee its output exclusively.

2.1.3 Value Co-creation

In traditional contexts, value creation is a process focused within the company and considers value as a quid to be transferred to the customer. The classical managerial approach to value creation is based on the model of the Porter's value chain (Porter 1985). According to this model, the value created by a company is the result of the interaction of nine characteristics divided into two categories: primary activities and support activities.

Primary activities: these activities concern the physical creation of the product. They are as follows:

- internal logistics: activities associated with the receipt, storage and distribution of the raw materials necessary for the manufacture of the product including treatment, storage, inventory, picking and sorting;
- operations: activities associated with the physical transformation of inputs into the final product and among which we include machining, packaging, finalization of pre-assembled, quality control and testing;
- external logistics: activities related to the collection, storage and distribution of finished products such as loading, unloading and transport;
- marketing and sales: set of activities aimed at supporting the sale of the product such as advertising, promotions, pricing, maintenance and structuring of relations;
- services: activities related to the support of the value of the product such as installation and training.

Support activities: these activities support those seen previously, guaranteeing all that is necessary for them to function at their best. They are as follows:

- procurement: purchase of the inputs necessary for manufacturing such as instrumentation, raw materials, component, pre-assembled and consumables;
- research and development: directed toward the innovation, introduction and improvement of products, services and processes;
- human resources management: activities related to human capital management, for example, selections, recruitment, training and remuneration;
- business infrastructure: high-level activities such as financial management, demand planning and general management.

PSS logics considers, the customer no longer as the point of arrival of the value creation process (as in the Porter's value chain illustrated here), but directly involved in its creation, moving towards an approach known as co-creation (Fig. 2.4). When we talk about co-creation, we mean the progressive involvement of the client in the value creation process. This type of relationship allows companies to review this strategic process and thus achieve new competitive advantages.

Concepts of co-creation are as follows:

- Company and customer create value together;
- The customer co-constructs the service to make it fit his/her needs;
- Problems are defined and resolved together;

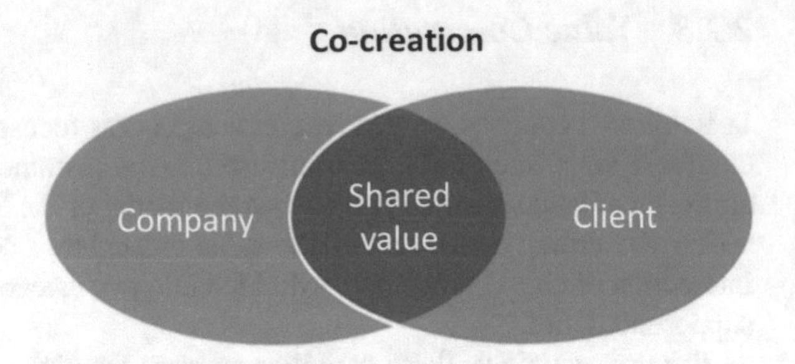

Fig. 2.4 The concept of value co-creation between the company and the client

- Co-creation of innovation;
- Development of nuances shifts the centre of gravity from consumption to customer experience.

How are the premises for co-creation built? First of all, we need to construct a structure of interaction between customer and company (Prahalad and Ramaswamy 2004). At the base of this structure, we find four elements (Fig. 2.5):

- dialogue: the possibility of having a constructive and continuous dialogue with the partner to better communicate needs and constraints;
- access: access to the infrastructure of the respective partner in order to acquire know-how regarding its skills and needs;
- risk–benefits: explanation of risk factors and benefits and how they should be broken down;
- transparency: transparency on the resources used and on the strategies used to coordinate the activities.

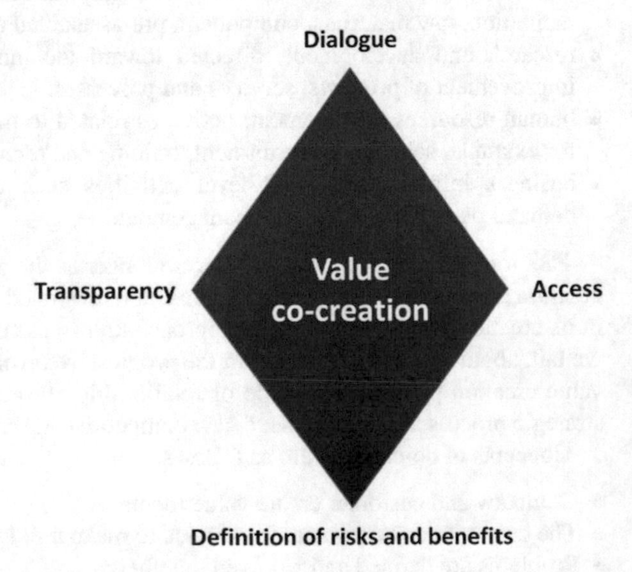

Fig. 2.5 Elements of value co-creation

In the co-creation phase, the customer is not just a simple external innovator, or a simple consultant, but becomes a partner who shares risks with the provider and shares access to his assets for reach the goal (Prahalad and Ramaswamy 2004).

2.1.4 Accessing the Value: From Ownership Towards Use

If traditionally the change in the ownership of the product is the purchase, in the PSS determining the owner of the asset is not always immediate. The customer does not buy the product, but the performance (Markeset and Kumar 2005) and its ownership depend on the type of contract. In this regard, it is possible to refer to the framework, already presented in Chap. 1 (Fig. 1.3), to examine in more detail how and in what phase of the product life cycle the property ownership can be shared.

The framework proposed in Fig. 2.6 describes property management and five most frequent concrete situations regarding the implementation of PSS (Lay et al. 2009).

The "ownership during the phase of use" parameter defines which of the two counterparties has property rights over the asset and related equipment during the period covered by the contractual terms. At the end of the agreed time frame, the provider can either hold the product or sell it to the customer at the market price. Between these two "extreme" options, there is a range of possible situations. Other players, such as banks, could buy the asset and lease it to the customer or the same provider, or the customer can choose to establish a cooperation with a bank or another external player to establish a joint venture. Venture aimed at purchasing the product. Naturally, the financial aspects of the various configurations described above must be carefully evaluated.

"Ownership after phase of use" refers to the parameter with which the property right of the product is specified after the end of its operating life. Usually, two possible options are configured: depending on who was previously the owner of the asset, the whole package can either remain the property of the supplier, or be resold to the provider that deals with the operations of updating or recycling the goods. In the case of a joint venture or leasing bank, the asset can be sold to the customer who assumes the management charges. In a nutshell, this parameter can be understood as an indicator used to understand who is responsible for the recycling of the product and its components.

Characteristic Features		Options			
Ownership	during phase of use	Equipment producer	Leasing bank	Operating joint venture	Customer
	after phase of use	Equipment producer	Leasing bank	Operating joint venture	Customer

Fig. 2.6 Ownership visualization scheme (Lay et al. 2009)

Characteristic Features		Options			
Ownership	during phase of use	Equipment producer	Leasing bank	Operating joint venture	Customer
	after phase of use	Equipment producer	Leasing bank	Operating joint venture	Customer

Fig. 2.7 Ownership: type 1 (Lay et al. 2009)

Characteristic Features		Options			
Ownership	during phase of use	Equipment producer	Leasing bank	Operating joint venture	Customer
	after phase of use	Equipment producer	Leasing bank	Operating joint venture	Customer

Fig. 2.8 Ownership: type 2 (Lay et al. 2009)

Five configurations of use and distribution of PSS possession consequently emerge and are characterized by different options of management and distribution of the property during the product life cycle:

Type 1: this category (Fig. 2.7), considers those experiences similar to the traditional business model that allows the customer to use a certain asset or machinery in the face of a payment without it passing ownership. This model can be seen as an evolution of the classic formula of the rental, where the payment is made on the basis of the number of accesses or the amount of transactions processed with the asset contractualized. If at the end of the contract the customer does not buy the goods, the provider may be interested in re-inserting the product (which often has not yet reached the end of its life cycle) in new production contests.

Type 2: the second type (Fig. 2.8) does not focus on financial aspects, but rather on operations. In this type, the personnel assigned to the tasks provided by the product and related to its maintenance are not dependent on the customer, but are in charge of the provider. With this formula, the customer acquires the goods from the supplier or receives it in leasing and, then, requests its exclusive installation in its plant. This situation is typical in contexts where the client is deprived of the adequate human resources necessary to use highly technological products at the best.

Type 3: the third type of products (Fig. 2.9) is a combination of the first and second types described above with orientations on financial and operational aspects. The supplier retains ownership over the asset, uses the equipment in the exclusive customer plant and employs personnel for operational and maintenance activities. The provider is paid according to use or based on the parts produced using the equipment that makes up the PSS supply.

Type 4: the fourth type (Fig. 2.10) considers that the ownership of the asset is still bound to the supplier and there is a strong similarity with the type of PSS seen in point 3. The substantial difference lies mainly in the location of the production equipment. The provider installs the equipment inside the facility or directly next to the customer's plant that is expected to serve and produces the components required

Characteristic Features		Options			
Ownership	during phase of use	Equipment producer	Leasing bank	Operating joint venture	Customer
	after phase of use	Equipment producer	Leasing bank	Operating joint venture	Customer

Fig. 2.9 Ownership: type 3 (Lay et al. 2009)

Characteristic Features		Options			
Ownership	during phase of use	Equipment producer	Leasing bank	Operating joint venture	Customer
	after phase of use	Equipment producer	Leasing bank	Operating joint venture	Customer

Fig. 2.10 Ownership: type 4 (Lay et al. 2009)

Characteristic Features		Options			
Ownership	during phase of use	Equipment producer	Leasing bank	Operating joint venture	Customer
	after phase of use	Equipment producer	Leasing bank	Operating joint venture	Customer

Fig. 2.11 Ownership: type 5 (Lay et al. 2009)

to fulfil the customer's orders. In this way, the provider is able to cope with customer demand peaks or offers a production capacity buffer in the event of malfunctions or breakages. Also, in this case, the personnel assigned to the operational and maintenance activities is in force of the provider, while the payment is made for units of products processed with the equipment made available.

Type 5: the last type of PSS (Fig. 2.11) is characterized by the involvement of the third party. It then takes the form of an operating joint venture or there is a contractor capable of catching up with the risks associated with the ownership of the asset investing in the purchase of the asset and using it on behalf of the client who may be considered as a partner. At the end of the supply contract, the asset becomes property of the customer.

At this point, it is natural to wonder why the provider and its customers should abandon a more traditional business model in favour of those described above. Looking at the benefits of each type, in the second type the substantial advantage consists of the greater ability of the provider to use its equipment. For the other types analysed, it is necessary to refer to the following statement: "the right of ownership in an asset is understood as the right to use the latter, to change its form and substance and to transfer its rights entirely or some parts thereof" (Furubotn and Richter 1998). In the traditional business model, these rights are ceded in the moment of sale. The different divisions of ownership into PSSs allow economies of scale to be obtained and information asymmetries to be reduced (Morey and Pacheco 2003). The first type of contracts, based on an evolution of the concept of rent, provides that the right of use is disconnected from the right of possession. This implies, from a customer point of view, that fixed costs become variable and that the real cost of use is evident (Hockerts 2008). From the supplier's point of view, however, it no longer makes sense to focus on aspects related to the sale of the product and related equipment when the payment is made for each unit of product processed with the equipment provided. Consequently, given the interest of the provider in recovering the good at the end of the contract, the duration of the supplied equipment needs to be expanded.

In the traditional models the customer had a poor knowledge of information about the characteristics of the supply and the relative modes of use that guarantee its operation. In order to balance this asymmetry and avoid being damaged by the opportunistic conduct of the provider, considerable financial resources are required. On the other hand, the manufacturer has full knowledge of its products and its potential so if the ownership of the equipment is not transferred to the customer but is maintained by the supplier, the customer does not have to bear the efforts to fill information asymmetry. In types 3, 4 and 5, the manufacturer is responsible for the use of machinery and can use his experience to achieve economies of scale.

2.2 The Key Issue in the Customer Management

2.2.1 Improved Relationships with Customers

In PSS, unlike traditional settings, customer relationships are critical success factors (Galbraith 2002; Tukker 2004; Gebauer et al. 2005; Davies et al. 2007; Kindström 2010; Reim et al. 2015). It is important to define the type of interaction that must be established with customers in order to transmit value and maintain it during the life cycle of the product (Meier et al. 2010; Barquet et al. 2013; Liu et al 2014). The increase in interactions with the client is a signal of evolution of the relationship towards a servitization logic (Azarenko et al. 2009). This also includes the definition of the ways in which information sharing should take place (Windahl and Lakemond 2010; Reim et al. 2015). Customer relationship management is strictly related to the generation of added value through direct connections and intensified contacts with the client (Mont 2004). This implies that relationships with customers are structured and long-term, as opposed to short-term treatment of the "product sale" context (Mont 2004; Williams 2006). The relational course is undertaken by establishing and constructing operational intersections, exchange of information, legal contracts and defining cooperation rules (Matthyssens and Vandenbempt 2010).

2.2.2 *Customer Interaction*

In PSS, a close relationship and improved interaction between company and customers are the basis for the success of the development and management of the solutions offered (Galbraith 2002; Davies et al. 2007; Cova and Salle 2008), allowing the mutual creation of value through the co-creation scheme. In fact, the success of value co-creation is heavily based on the involvement and the client's efforts (Sheth and Uslay 2007).

Customer participation in design, production, sales and delivery are typical of PSS (Kindström and Kowalkowski 2009). This implies that sporadic interactions become continuous over time and require a contractual structuring that will be further investigated. The boundaries between customer and supplier are therefore permeable to information and experiences, favouring the osmosis of knowledge and skills that enriches both. Given this strategic aspect, if the interactions are not managed carefully, the process of enrichment of the solutions offered cannot be unlocked, leading to the failure of the customer experience. It is possible to design the interaction with the client by analysing four aspects (Fig. 2.12):

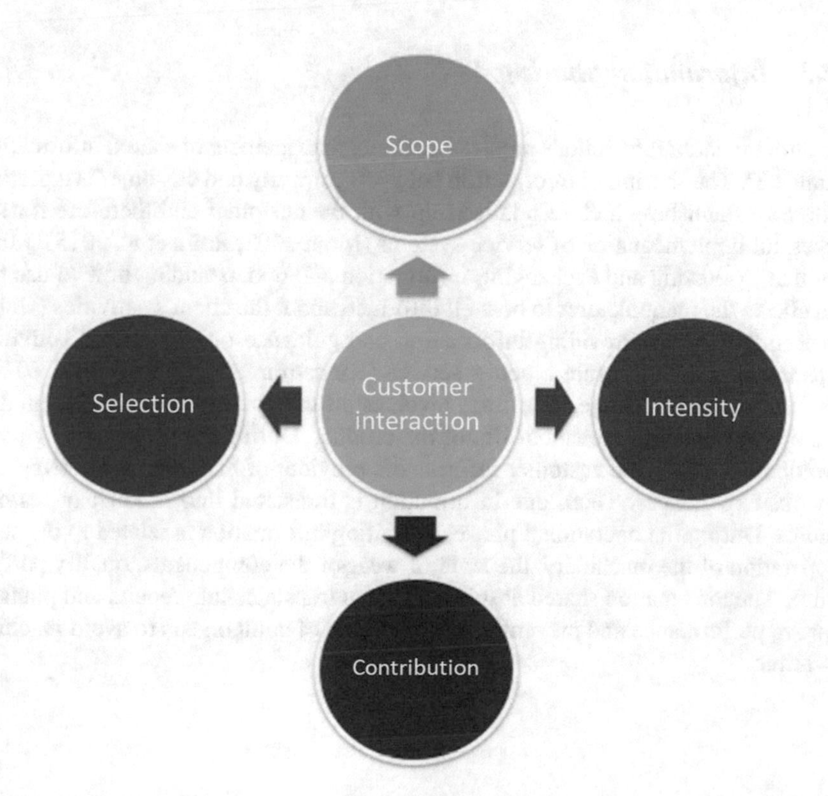

Fig. 2.12 Key characteristics of customer interaction

- Scope: the scope refers to the propensity of the provider to involve the clients along all the phases of the development of the PSS. During development stage, customer activities are an integral part of the value co-creation process and the provider must interact more closely with the customer. Parallel to the life-cycle phases of the product, also customer–supplier relationships are developed. The life cycle of the product consists of conceptualization, design, production, testing, installation, use and maintenance, while the *cycle of interaction* passes through phases related to access to information, diagnosis and delivery until follow-up.
- Intensity: intensity refers to three levels of customer involvement in the development process: "FOR the customer", "WITH the customer" and "FROM the customer". The first level implies the highest responsibility for the provider, the second one the cooperation between the two and the last one the maximum responsibility on the client.
- Contribution: the set of contributions made by the customer during the co-creation phase.
- Selection: the phase consists in selecting the contributions that the client can bring during the co-creation phase.

2.2.3 Information Sharing

The enrichment of interactions requires a correct management of shared information (Table 2.3). The sharing of information between company and customer is a prerequisite for establishing a close relationship with the customer and therefore for the successful implementation of service systems (Mont 2002; Reim et al. 2015). More than that, collecting and exchanging information and understanding how to use the data allows the manufacturer to be well informed about the client's activities (Ulaga and Reinartz 2011). Providing information and guidance on operational activities helps the supplier to ensure a better service (Kindström and Kowalkowski 2014). The information exchange, therefore, involves all the phases, from the design and development phase to the end of life of the product. During the product design and development phase, the customer informs the provider of his needs, objectives and previous experiences. Then, this information is translated into product or service features. During the operational phase, prevailing information is related to the state of operation of the machinery, the state of wear of the components, quality performance. The information shared at this stage then translates into repairs and plans to improve performance and preventive maintenance of components to avoid machine downtime.

Table 2.3 Classification of shared information

Life-cycle phase	Shared information	Related aspects
Product development	• Needs • Experiences • Design capability • Information systems for design and engineering	• Product specifications
Operational phase	• Data related to the solution's operation • Wear condition • Availability	• Preventive maintenance • Repairs • Performance improvement plans

2.2.4 Sales Channels' Effect in Value Communication

Understanding how value is transmitted to the customer is fundamental, but companies should also rethink how to create awareness on the new service offered and how to communicate the added value (Reim et al. 2015). In order to allow the transition from product-centric sales to PSS logics, the sales areas should make the PSS option more attractive than the traditional basic products (Tukker and Tischner 2006) and to do so require adequate preparation in order to "sell the idea" with targeted marketing campaigns. The search for new ways to transmit the value of the PSS involves a new definition of the pre- and post-sales channels, through an internalization or outsourcing of specific resources in order to develop or acquire new skills (Storbacka 2011; Kindström and Kowalkowski 2014).

Sales Channel Configuration

Sales channels must be able to create customer awareness and facilitate evaluation of the offer. The personnel involved in these activities must therefore be accredited in terms of reliability, knowledge of the PSS and must become a resource to create added value for customer. So, sales forces should change their sales strategies (Kindström et al. 2015). Sales parameters must be focused on the perception of value and not on internal costs. Given the very complex and personalized nature of the PSS, the most suitable sales channel is the direct one. Relying on third parties would be complicated and difficult to implement given the nature of the information that is shared throughout the product life cycle. The sales channel therefore adapts to the reality of the PSS context. Usually, when a company distributes solutions in the B2B context, the practice is to use direct sales channels, while in the B2C context indirect channels are used, also considering the lower complexity of the solutions offered (Nordin 2005) (Figs. 2.13 and 2.14).

After-Sales Channel

Once the channels concerning the distribution of the asset have been designed, it is necessary to concentrate on those concerning the after-sales services. The management of the field service network is a key component in the success of PSS delivery. This includes a, for example, repair or maintenance of the product or its components

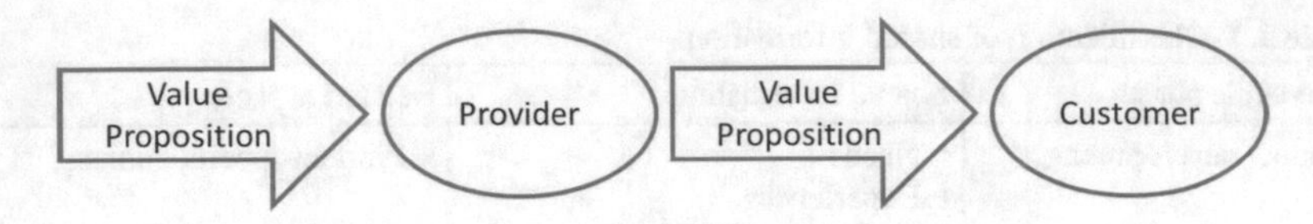

Fig. 2.13 Direct distribution channel (B2B)

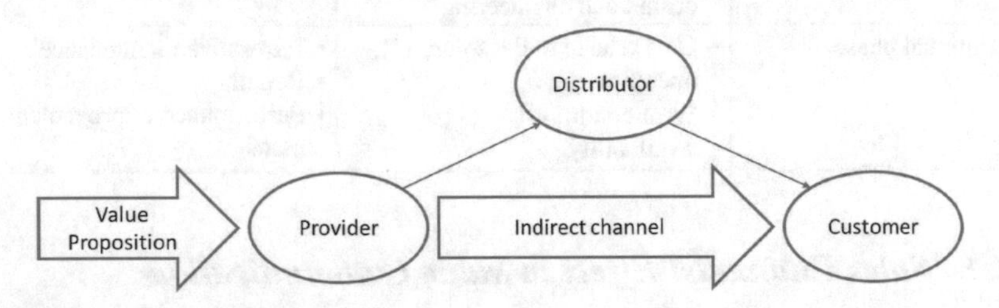

Fig. 2.14 Indirect distribution channel (B2C)

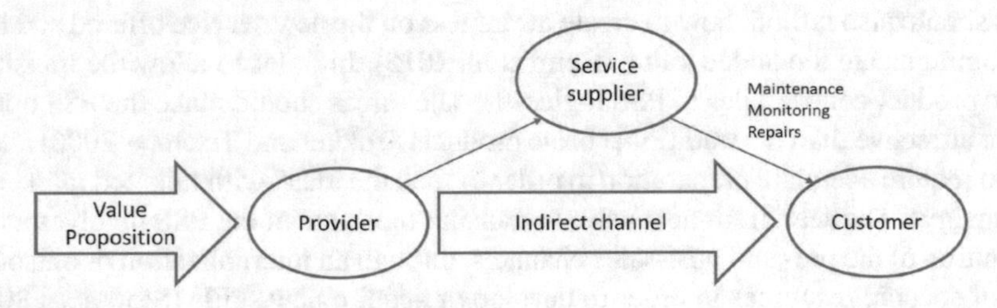

Fig. 2.15 Indirect channel, service supply (B2C)

at the end of the life cycle before and after disposal or recycling of materials. If you talk about activities that directly support operations, the technicians take great care. The technicians, in fact, are often in contact with the customer and the latter tends to be confident with them (Ulaga and Reinartz 2011). For this reason, in the PSS business models the after-sales channel should be highly integrated with the sales channel dedicating to the customer a unique and specific line of communication on which the entire customer–supplier relationship is based. Coherently with the specificity of the communication channel, specific teams must be set up to manage customer problems (Gebauer and Kowalkowski 2012; Kindström et al. 2015). To understand how the after-sales channel should be managed, the complexity of the service and its intensity must be assessed. Usually, when the service is very complex in terms of know-how and must be provided with extreme frequency, the practice is to make use of a subcontractor who works on behalf of the provider, while if the service component is relatively simple to manage, or should be provided with low frequency, the provider should evaluate the internalization of this activity to maintain a direct channel with the customer (Nordin 2005) (Fig. 2.15).

2.2.5 Different Contractual Models

The introduction of the PSS requires the supplier to move to a new pricing discipline. The traditional cost-plus approach, in which the price was obtained by summing the costs of production, design and development and a margin which constituted the producer's profit, is replaced by a value-based logic (Rapaccini and Visintin 2015); in this case, it is more difficult to understand if and how the product will cover the costs incurred and risks and potentialities are difficult to predict, but the new pricing models guarantee profitability because very often the value that the customer is willing to pay is higher than that obtained with cost-plus methods (Oliva and Kallenberg 2003; Tukker and Tischner 2006; Neely 2009). Ownership of the asset is one of the most critical aspects of the contract: as seen above, it can remain in the hands of the supplier or be transferred to the customer. If it is not transferred to the customer, the rights concerning all the activities related to it must be carefully established in order to avoid unpleasant disputes during the use phases (Richter et al. 2010) and it is crucial to define reliable outputs to be included in contractual terms (Bonnemeier et al. 2010). There is a need to turn the offer into terms and specifications such as to describe what and how it is distributed to the parties of the contract (Azarenko et al. 2009; Meier et al. 2010; Reim et al. 2015). The contracts are complex and concern the procedures and penalties that take place in the case of non-satisfaction of the clauses. Also, in this case, it is possible to use an approach that analyses the nature of contracts according to their context of use (Reim et al. 2015). A contract PSS should be built with a view to shed light on all aspects concerning the rights and duties of each party.

Many supply contracts are extremely complex and their terms must be adapted taking into account the context of the PSS. This complexity varies according to the quantity of specified regulations, so it will vary according to the business model linked to the PSS. Contrary to complexity, the level of contract formalization indicates how much it should be readjusted to each new customer. Very formal contracts tend to be less complex, since they have to adapt to a vast number of contexts (Reim et al. 2015). Long duration is an unavoidable factor in PSS contracts, so companies should establish an appropriate balance between the interests of the parties. According to Richter and Steven (2009), the contractual phase plays a key role in the definition of the business model: its formulation has a greater impact on profits than the choice of the business model with which to operate. In order to maximize the value generated, it is essential to align the characteristics of the business model with the contractual terms regarding the aspects of liability and risk representation. More in detail, there are three key aspects to be carefully taken into account: (1) *responsibility and terms of the agreement*, (2) *complexity and formalization* and (3) *level of risk*.

1. *Responsibility and terms of the agreement:* concern how the tasks are divided between the parts of the contact and which specifications are necessary to clarify rights charges from a purely legal perspective. In product-oriented types, the customer is the owner of the product and the only responsibility of the provider regards the services related to the product. This means that the contract must

establish and define the level of service provided and the outputs. With a supply contract, the tasks to be performed and the time frames to complete the activities must be included. It is equally important to agree on payment terms and how additional costs are credited in case of repairs (Azarenko et al. 2009). The contract should concern the management of shared information (Schuh et al. 2011). In use-oriented contexts, certain terms such as availability, price, control over machinery and responsibilities for losses caused by periods of non-availability must be reported in the contract. In this case, since the ownership of the product is not transferred to the customer, the decision rights must be allocated carefully (Richter et al. 2010). The client's responsibility is greater in the use-oriented context rather than the product-oriented one, but reaches its maximum in result-oriented realities because the provider has complete responsibility for ensuring the result (Meier et al. 2010). As the level of responsibility increases, the terms of the agreement become extremely important. This not only leads to increased responsibility, but also a great need to share information. Often, however, information can be sensitive, so there is a need to agree on what information to build interchange.

2. *Complexity and formalization of the contract:* formalization is higher in contracts for product-oriented because this type of offer solutions is very standardized, and this makes very similar contracts possible in different contexts. The lowest level of formalization is expected to be found in result-oriented models because they offer unique and unrepeatable solutions to each individual customer. The complexity increases with the increasing responsibility of the provider. Agreeing on the services provided when working in a product-oriented context is not very complicated and both parties must check whether the shrewd are respected or not. The level of complexity is maximum in result-oriented contexts because the result must be guaranteed according to well-defined specifications. Moreover, as the customer–supplier relationship grows, the complexity of the agreements also increases. In these cases, it may be useful to make use of several parallel contracts (Azarenko et al. 2009).
3. *Level of risk:* usually, the level of risk increases when the provider moves from a product-oriented to a results-based model, but this is not necessarily valid for all types of PSS. The provider could see a way to secure premium incentives when taking a major risk share. In product-oriented contexts, risks are mainly linked to situations where more resources are needed to meet the terms of the contract, which would oblige the provider to review its operations. However, even a customer's averse behaviour is also a risk that can be mitigated through terms added to the agreement (for example, revocation of the guarantee when the customer does not meet the terms of the contract) (Azarenko et al. 2009). The risk of incorrect behaviour of the customer increases in the case of use-oriented models because the ownership of the product remains in the hands of the provider. This makes it necessary to agree on the decision rights and what costs will be linked to the use will be discharged on the customer (Reim et al. 2015). For suppliers, the main incentive for this type of contract is the higher revenue expected from the service offered. In result-oriented contexts, in

Table 2.4 Schematization of contracts characteristics

PSS category	Liability and terms of the agreement	Formalization and complexity	Risk component
Product-oriented	• Charges for services • Agreement on tasks, payments and information management	• High formalization • Low complexity	• Low risk • Adverse behaviour
Use-oriented	• Charges concerning availability • Definition of the level of availability and monitoring activities	• Average formalization • Average complexity	• Average risk • Adverse behaviour
Result-oriented	• Charges concerning performance	• Low formalization • High complexity	• High risk • More freedom for provider

which the contract is based on the guarantee of certain performance, the risks are mainly based on the achievement of the patented results. In this case, the entire responsibility falls on the provider and usually this type of solution is proposed only by a limited number of entities capable of taking on such a risk at a high premium. The client, for his part, benefits from the reduction of the efforts necessary to achieve certain results.

Table 2.4 summarizes the different contractual models of the three PSS categories

Seven Key Facts

- The shift towards servitization brings significant changes in the role of customers and their involvement from the design to the delivery of the offering.
- The value proposition at the core of servitization can be declined according to customer value, value co-creation, product ownership and service offering.
- The value for customer can consist of tangible and intangible elements, i.e. performance, customization, cost reduction, risk reduction, usability and contract flexibility.
- The service offering is one of main distinctive elements of a PSS: there can be services that support the product or services that support customer activities, and the value logic can be focused on inputs or either on outputs.
- Value co-creation is a central element in the value proposition of servitization: the customer is directly involved in value creation process.

- Unlike traditional selling, servitization does not (always) involve shifts in product ownership: the customer pays for product's performance and not for possession.
- Key element for the success of servitization and PSS is a mindful management of relationship with customers and their involvement: this might imply a redefinition of channels, interactions, and contracts.

References

T. Alonso-Rasgado, G. Thompson, B.O. Elfström, The design of functional (total care) products. J. Eng. Des. **15**(6), 515–540 (2004)

A. Azarenko, R. Roy, E. Shehab, A. Tiwari, Technical product-service systems: Some implications for the machine tool industry. J. Manufact. Technol. Manage. **20**(5), 700–722 (2009)

A.P.B. Barquet, M.G. de Oliveira, C.R. Amigo, V.P. Cunha, H. Rozenfeld, Employing the business model concept to support the adoption of product-service systems (PSS). Ind. Mark. Manage. **42**(5), 693–704 (2013)

S. Bonnemeier, F. Burianek, R. Reichwald, Revenue models for integrated customer solutions: concept and organizational implementation. J. Revenue Pricing Manage. **9**(3), 228–238 (2010)

B. Cova, R. Salle, Marketing solutions in accordance with the SD logic: co-creating value with customer network actors. Ind. Mark. Manage. **37**(3), 270–277 (2008)

A. Davies, Moving base into high-value integrated solutions: a value stream approach. Ind. Corp. Change **13**(5), 727–756 (2004)

A. Davies, T. Brady, M. Hobday, Organizing for solutions: systems seller vs. Systems Integrator. Ind. Mark. Manage. **36**(2), 183–193 (2007)

B.K. Fishbein, L.S. Mcgarry, P.S. Dillon, *Leasing: A step Toward Producer Responsibility* (INFORM, New York, 2000)

E.G. Furubotn, R. Richter, *Institutions and Economic Theory: The contribution of the New Institutional Economics* (University of Michigan Press, Ann Arbor, IL, 1998)

J. Galbraith, *Designing Organizations: An Executive Guide to Strategy, Structure and Process* (Jossey-Bass, San Francisco, CA, 2002)

H. Gebauer, E. Fleisch, T. Friedli, Overcoming the service paradox in manufacturing companies. Eur. Manage. J. **23**(1), 14–26 (2005)

H. Gebauer, C. Kowalkowski, Customer-focused and service focused orientation in organizational structures. J. Bus. Ind. Mark. **27**(7), 527–537 (2012)

C. Grönroos, A service perspective on business relationships: the value creation, interaction and marketing interface. Ind. Mark. Manage. **40**(2), 240–247 (2011)

K. Hockerts, Property rights as a predictor for the ecoefficiency of product-service systems. *Working Paper No. 02–2008, CBS Center for Corporate Social Responsibility, Frederiksberg*, 2008

O. Isaksson, T.C. Larsson, A. Öhrwall Rönnbäck, Development of product-service systems: challenges and opportunities for the manufacturing firm. J. Eng. Des. **20**(4), 329–348 (2009)

D. Kindström, C. Kowalkowski, Development of industrial service offerings: a process framework. J. Serv. Manage. **20**(2), 156–172 (2009)

D. Kindström, Towards a service-based business model—key aspects for future competitive advantage. Eur. Manag. J. **28**(6), 479–490 (2010)

D. Kindström, C. Kowalkowski, Service innovation in product centric firms: a multidimensional business model perspective. J. Bus. Ind. Mark. **29**(2), 96–111 (2014)

D. Kindström, C. Kowalkowski, T.B. Alejandro, Adding services to product-based portfolios: an exploration of the implications for the sales function. J. Serv. Manage. **26**(3), 372–393 (2015)

S. Kujala, K. Artto, P. Aaltonen, V. Turkulainen, Business models in project-based firms—towards a typology of solution-specific business models. Int. J. Project Manage. **28**(2), 96–106 (2010)

G. Lay, M. Schroeter, s Biege, Service-based business concepts: a typology for business-to-business markets. Eur. Manag. J. **27**(6), 442–455 (2009)

C.H. Liu, M.-C. Chen, Y.-H. Tu, C.-C. Wang, Constructing a sustainable service business model: an S-D logic-based integrated product service system. Int. J. Phys. Distrib. Logistics Manage. **44**(1–2), 80–97 (2014)

T. Markeset, U. Kumar, Product support strategy: conventional versus functional products. J. Qual. Maintenance Eng. **11**(1), 53–67 (2005)

V. Mathieu, Service strategies within the manufacturing sector: benefits, costs and partnership. Int. J. Serv. Ind. Manag. **12**(5), 451–475 (2001)

P. Matthyssens, K. Vandenbempt, Service addition as business market strategy: identification of transition trajectories. J. Serv. Manage. **21**(5), 693–714 (2010)

H. Meier, R. Roy, G. Seliger, Industrial product—service system—IPS2. CIRP Ann-Manufact. Technol. **59**, 607–627 (2010)

O. Mont, Clarifying the concept of product-service system. J. Clean. Prod. **10**, 237–245 (2002)

O. Mont, Product-service system: panacea or myth? (doctoral thesis), Retrieved from the National Library of Sweden database. 91-88902-33-1 (2004)

E. Morey, D. Pacheco, Product-service systems: exploring the potential for economic and environmental efficiency. *Working paper, ECON 4545*, University of Colorado, 2003

M. Morris, M. Schindehutte, J. Allen, The entrepreneur's business model: toward a unified perspective. J. Bus. Res. **58**(6), 726–735 (2005)

A. Neely, Exploring the financial consequences of the servitization of manufacturing. Oper. Manage. Res. **1**, 103–118 (2009)

I.C. Ng, R. Maull, N. Yip, Outcome-based contracts as a driver for systems thinking and service-dominant logic in service science: evidence from the defence industry. Eur. Manag. J. **27**(6), 377–387 (2009)

F. Nordin, Searching for the optimum product service distribution channel: examining the actions of five industrial firms. Int. J. Phys. Distrib. Logistics Manage. **35**(8), 576–594 (2005)

R. Oliva, R. Kallenberg, Managing the transition from products to services. Int. J. Serv. Ind. Manag. **14**(2), 160–172 (2003)

A. Osterwalder, Y. Pigneur, *Business Model Generation: A Handbook for Visionaries, Game Changers, and Challengers* (John Wiley and Sons, Hoboken, NJ, 2010)

K.S. Pawar, A. Beltagui, J.C. Riedel, The PSO triangle: designing product, service and organisation to create value. Int. J. Oper. Prod. Manage. **29**(5), 468–493 (2009)

A. Payne, S. Holt, Diagnosing customer value: integrating the value process and relationship marketing. Br. J. Manage. **12**(2), 159–182 (2001)

C.K. Prahalad, V. Ramaswamy, Co-creation experiences: the next practice in value creation. J. Interact. Mark. **18**(3), 5–14 (2004)

M.E. Porter, *The Competitive Advantage: Creating and Sustaining Superior Performance* (Free Press, NY, 1985)

M. Rapaccini, F. Visintin, Devising hybrid solutions: an exploratory framework. Prod. Plann. Control **26**(8), 654–672 (2015)

W. Reim, V. Parida, D. Örtqvist, Product-service systems (PSS) business models and tactics—a systematic literature review. J. Clean. Prod. **97**, 61–75 (2015)

A. Richter, M. Steven, On the relation between industrial product service systems and business models. Oper. Res. Proc. **2008**, 97–102 (2009)

A. Richter, T. Sadek, M. Steven, Flexibility in industrial product-service systems and use-oriented business models. CIRP J. Manufact. Sci. Technol. **3**, 128–134 (2010)

G. Schuh, W. Boos, S. Kozielski, Lifecycle cost-orientated service models for tool and die companies. In *Proceedings of the 1st CIRP Industrial Product-Service Systems (IPS2) Conference, 249*, Cranfield University Press, 2009

J.N. Sheth, C. Uslay, Implications of the revised definition of marketing: from exchange to value creation. J. Publ. Policy Marketing **26**(2), 302–307 (2007)

K. Storbacka, A solution business model: capabilities and management practices for integrated solutions. Ind. Mark. Manage. **40**(5), 699–711 (2011)

L. Smith, I. Ng, R. Maull, The three value proposition cycles of equipment-based service. Prod. Plann. Control **23**(7), 553–570 (2012)

A.R. Tan, D. Matzen, T. McAloone, S. Evans, Strategies for designing and developing services for manufacturing firms. CIRP J. Manufact. Sci. Technol. **3**(2), 90–97 (2010)

A. Tukker, Eight types of product-service system: eight ways to sustainability? Experience from SusProNet. Bus. Strategy Environ. **13**, 246–260 (2004)

A. Tukker, U. Tischner, Product-service as a research field: past, present and future. Reflection from a decade of research. J. Clean. Prod. **14**, 1552–1556 (2006)

W. Ulaga, W.J. Reinartz, Hybrid offerings: how manufacturing firms combine goods and services successfully. J. Mark. **75**(6), 5–23 (2011)

S.L. Vargo, R.F. Lusch, Evolving to a new dominant logic for marketing. J. Mark. **68**(1), 1–17 (2004)

A. Williams, Product-service systems in the automotive industry: the case of micro-factory retailing. J. Clean. Prod. **14**, 172–184 (2006)

C. Windahl, N. Lakemond, Integrated solutions from a service centered perspective: applicability and limitations in the capital goods industry. Ind. Mark. Manage. **39**(8), 1278–1290 (2010)

Chapter 3
Product Service Systems' Competitive Markets

This chapter introduces three keystones of the competitive advantage given by the servitization strategy: sustainability, sharing economy/collaborative consumption and circular economy.

Subsequently, we analyse a key issue concerning whether PSS brings significant changes in the competitive structure of a market, and how it may affect in different ways B2B and B2C contexts. Since no specific trends seem to emerge in relationship with a particular category, the final aim of this chapter is presenting different examples of PSSs drawn from the most relevant industries impacted by servitization: manufacturing, sustainability driven and digital driven ones.

3.1 Contemporary Social and Economic Context

3.1.1 Sustainability

In the last 30 years, sustainability and its threefold impact on economic, environmental and social dimensions become a key point for managers seeking new chances and willing to explore new paths in the value creation process. Pressures determined by competitive contexts, national and international policies, together with the "green paradigm", also shaped the aspect of sustainability.

According to Elkington (2002), the process of sustainable development "involves the simultaneous pursuit of economic prosperity, environmental quality, and social equity", while an extension of this concept can be found in sustainable manufacturing, which "refers the creation of manufactured products that use processes that are non-polluting, conserve energy and natural resources, and are economically sound and safe for employees, communities, and consumers" (Khorram Niaki and Nonino 2018). In this context, the ability of addressing economic, environmental and social aspects through integration is a key aspect in the development of successful business models (Lozano 2008). Even if sustainability is acquiring a considerable impor-

A. Annarelli et al., *The Road to Servitization*,
https://doi.org/10.1007/978-3-030-12251-5_3

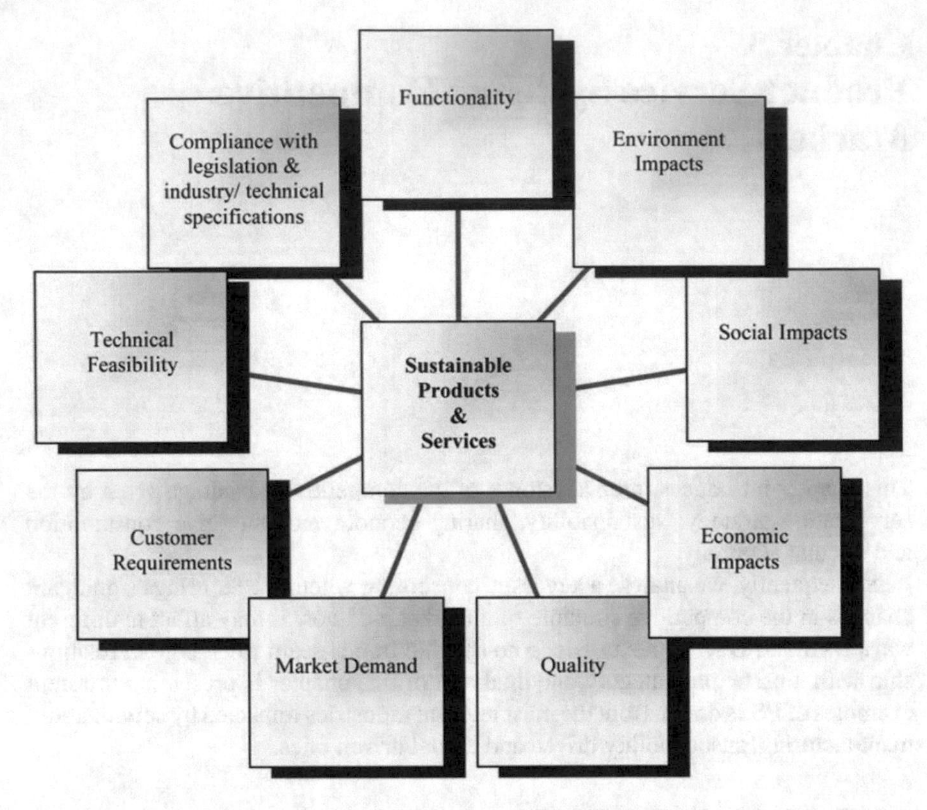

Fig. 3.1 Key criteria to optimize sustainability through PSS (Maxwell and van der Vorst 2003)

tance, there are, however, some uncertainties for future sustainable developments that remain (partially or completely) unaddressed like, for instance, issues related to fossil fuels, their unpredictability and fluctuations of prices, the usage in production processes of non-renewable inputs (raw materials characterized by scarcity), environmental pollution that continues having a rising trend and other issues such as space availability, landfill taxes and the need for more radical measures for environmental protection.

PSS can be a possible answer towards these sustainability concerns since it allows companies to address systematically and simultaneously all three dimensions of sustainability. Indeed, Maxwell and van der Vorst (2003) listed an exhaustive series of criteria to highlight this PSS's unique capability which are in particular *functionality*, *quality*, *technical feasibility*, *compliance with legislation* and others, as shown in Fig. 3.1.

Furthermore, Shokoyar et al. (2014) developed the concept of sustainable product service system, schematized in Fig. 3.2. With this term, authors labelled a specific type of PSS that is designed and implemented with the specific aim of achieving sustainable development during the product life cycle and its end of life.

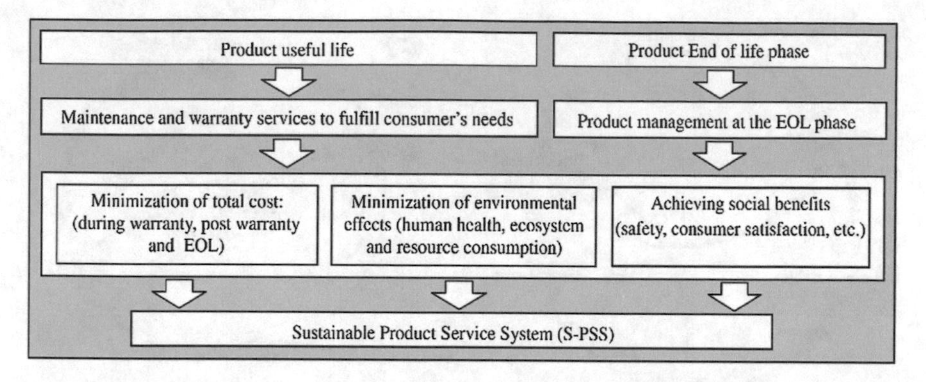

Fig. 3.2 The concept of sustainable product service system (Shokoyar et al. 2014)

Examples of organizations employing PSS offerings are the case of AMG (a company selling natural gas in Palermo, IT) offering a "solar heat service": "The consumer pays for receiving a service, comprehensive of final result, from installation to the thermal-energy meters and the transportation of methane to the boilers. It also granted the maintenance of the equipment" (Manzini and Vezzoli 2003). Thanks to this, "firms will have an incentive to prolong the service life of products, [...] to make them as cost- and material-efficient as possible, and to re-use parts as far as possible after the end of the product's life" (Tukker 2015). Furthermore, product design and manufacturing can no longer be the only source of competitive advantage and differentiation: product–service integrated solutions bring innovation potential, adding value to the total offering (Roy and Cheruvu 2009). This could be the simple case of extra services added to the product offering, with the aim of prolonging product life cycle and utility through time (for a more sustainable performance), while providing to customers a more satisfactory experience, worthy of extra revenue.

The above definitions and examples highlight PSS's ability in covering at the same time the three main dimensions of sustainability addressing social and environmental issues, while still presenting a meaningful value proposition, together with a high degree of flexibility offered, which is the main advantage behind the success of this business model (Barquet et al. 2013; Velamuri et al. 2013). In recent years, thanks to ICT advancement and digitalization, sustainability gained new life, with the rise of new concepts, i.e. circular economy (Witjes and Lozano 2016) and sharing economy (Bouncken and Reuschl 2018). Environmental (and social) concepts acquired through years a key importance in determining firms' performance and success over competitors (Kuijken et al. 2017). Among factors securing long-term competitiveness and operational success, the development of sustainable offerings plays a major role: reaching social and environmental achievements can have significant economy consequences (Miles et al. 2009; Patzelt and Shepherd 2011). Given this context, PSS can act as connecting point between environmental/social performance and economic needs of businesses, because it allows companies to meet evolved customers' needs with also a clear focus on sustainability: thanks to this, organizations can secure

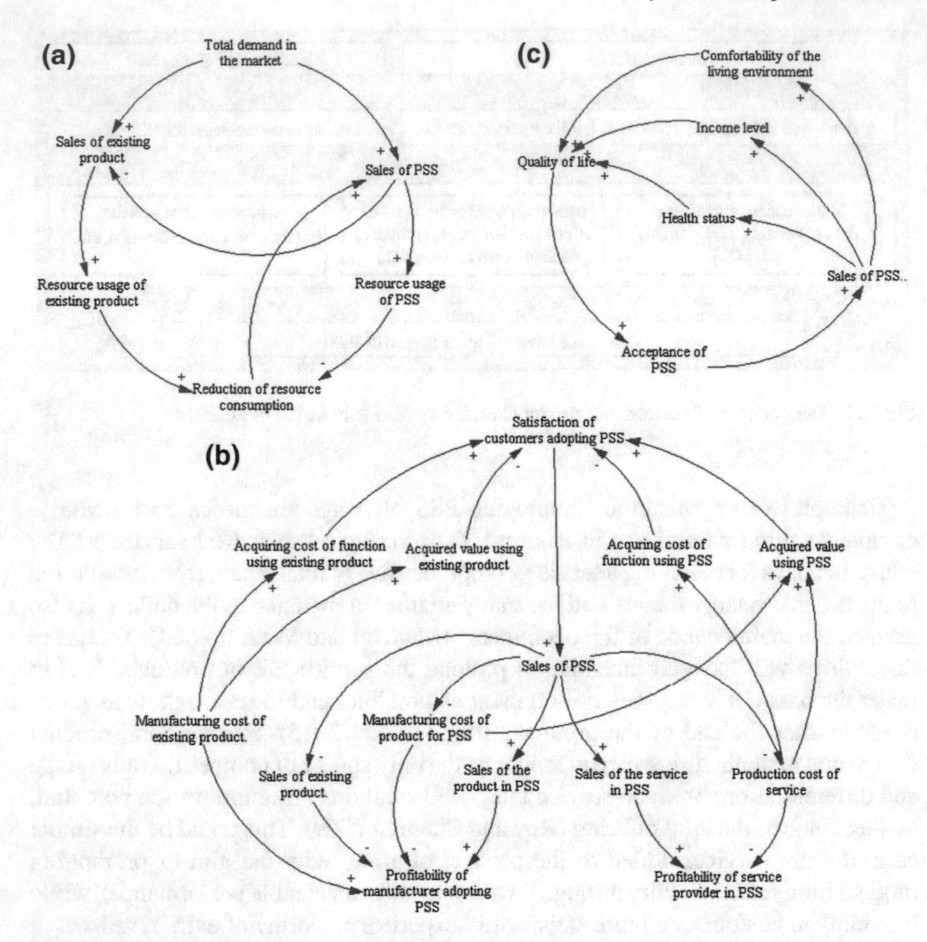

Fig. 3.3 Causal loop diagrams of environmental (**a**), economic (**b**) and social (**c**) dimensions of sustainability (Lee et al. 2012)

competitiveness and sustainability at the same time (Cook et al. 2006; Azarenko et al. 2009).

To give a graphic representation, Lee et al. (2012) adopted system dynamics to represent the impact of PSS on the three dimensions of sustainability (Fig. 3.3). In the case of social impact (Fig. 3.3c), for instance, there is a positive influence of quality of life on PSS acceptance, which, therefore, determines a rise in PSS sales, and these sales can impact different aspects of wellness and social dimension, resulting in an improved level of life quality, generating a positive cycle.

3.1.2 Sharing Economy and Collaborative Consumption

In recent years, collaborative-consumption-based businesses have become increasingly common. The concept of collaboration can be expressed in different forms: it can involve face-to-face interactions or it can be developed through the Internet, increasing peer-to-peer bonds. Many customers are offering access to products and services without the need for exclusive possession. The concept of collaborative consumption was developed by Rachel Botsman and Roo Rogers in the book "What's mine is yours: the rise of collaborative consumption" (2010). This concept was presented in the list of "Ten ideas that can change the world"; written by Time magazine (2010). What Rachel Botsman emphasized is that people have evolved from simply sharing information to connecting with others until they come to collaboration: there has been a transition from "I" to "We".

Furthermore, new network technologies (for example, smartphones and social networks) provide the efficiency and social glue that drive a critical mass of collaboration. A new belief in the relevance of the community is leading to a change in lifestyle; it is slowly moving from the era of unbridled consumerism to that of sharing and cooperation.

A complete definition of the phenomenon under analysis is as follows: "Collaborative consumption is a cultural and economic model based on access to goods rather than their exclusive possession; through technology and peer to peer, the movement reinvents the traditional concepts of sharing, borrowing, trading, renting, donating and exchanging" (Botsman and Rogers 2010). Therefore, the fundamental pillars of this phenomenon are possibility of reaching a critical mass, effective exploitation of goods that would otherwise be unusable, intelligent management of common goods and trust towards people who do not know each other directly.

The first aspect to consider for understanding collaborative consumption is the definition of the actors involved. There are two possible actors who take part in this process; the first is the "peer provider", the person or organization that provides the assets of sharing and the second is the "peer user" that is the person or organization that physically benefits from the available assets. Another preliminary step is the classification of the types of collaborative consumption based on the characteristics they possess. This is necessary because many businesses such as local exchange trading systems, social lending, car sharing, clothing swapping, peer-to-peer rentals and many others are collaborative consumption systems and should be organized into different categories according to pre-defined criteria, as for example the type of relationship/exchange between peers or the nature of underlying goods.

Together with the concept of collaborative consumption there is also that of collaborative lifestyle. This is mostly intangible in the sense that involves the exchange of intangible assets such as time, space and activities. More precisely, it concerns the sharing of workspaces, gardens and food; it also includes sites related to social loans and the rental of special housing.

Case Study

The Context of Sharing Economy

Case 1: BlaBlaCar

The history of BlaBlaCar develops in parallel in two European states. In France, BlaBlaCar was found in Paris in 2006 under the name of Covoiturage.fr and became established in just a few years as a new mode of transport for millions of French, especially young people. In Italy, in February 2010, postoinauto.it was born from the idea of a group of students who had experienced trips shared abroad, particularly in France and Germany. In March 2012, postoinauto.it became part of the now international network of Covoiturage.fr, which changes its name to BlaBlaCar. Following this important step, the postoinauto.it team decides to rename the site in BlaBlaCar.it, determining the effective union of the two platforms.

Once registered, the user can offer and/or request passages to other users in turn. It is required to simply choose the date, place of departure and place of arrival and the options will appear on the screen available. These are programmed shifts from city to city, trips that drivers would have anyway performed alone. Having free seats, the costs can be amortized by sharing one's own half with other users. While private transport within cities is very regulated, as far as long-distance travel is concerned, there are no particular rules to follow and those who offer a ride on BlaBlaCar do not do it for work.

Who are those who come into contact with the BlaBlaCar business? On the one hand, those who sign up for the platform with the purpose of providing one or more seats and, on the other hand, the people who use this service. On the one hand, there is the search for a possible monetary saving of those looking for a passage and, on the other, the possibility of sharing one or more seats, thus benefiting from both an economic and a social point of view. BlaBlaCar has two categories of customers: drivers and passengers. Drivers are people who own a car and have one or more seats available that they want to share with one or more passengers. The latter are all those who do not own their own cars, those who want a cheaper alternative to other forms of transport or those who simply want to travel and meet new people.

Users can register and create an account through the company's website or through their app. Once registration is complete, users can access all platform options, search for passes and passengers, evaluate other users and record payment details. The company provides support to its users through frequently asked questions and online forms. In addition, BlaBlaCar keeps its subscribers updated on platform developments through its blog and allows customers to interact with it through various social channels.

The platform allows customers to save on long-distance travel by sharing costs with one or more users. The BlaBlaCar community functionality makes driving a safer and more social experience as drivers are not forced to make long journeys on their own.

The company's rating system also allows users to view and evaluate potential travel partners before accepting it. Because security can be of particular importance to users who rely on travel-sharing services, BlaBlaCar offers an option for women only, allowing members to create travels where the driver and all passengers are women.

All BlaBlaCar journeys are insured for free, with travel quotas that do not affect drivers' insurance policies. In addition, BlaBlacar encourages a more environmental friendly approach by using parking spaces that would otherwise be empty.

The possibility to choose their own passengers and, on the other hand, their driver can become in some ways a negative aspect of the service: in this way, the most socially marginalized people will find it difficult to be accepted by other users.

BlaBlaCar operates a network of partners, including franchisees for marketing, travel agencies and dealers. All BlaBlaCar races in France and the United Kingdom are covered by AXA's travel-sharing insurance.

In addition, the company has started marketing partnerships with travel operators and markets. He recently collaborated with IRCTC, a subsidiary of Indian railways, to advertise its ride-sharing services to users who wish to book tickets for city-to-city journeys.

BlaBlaCar has also established a partnership with the European VINCI Autoroutes motorway concession operator, which now offers an automated toll payment for two trips each month booked on BlaBlaCar.

Decisions about BlaBlaCar's business decisions are influenced by its board of directors and all the various investors in the company.

The main resources of BlaBlaCar are its software platform, IT infrastructure and marketing partners. Furthermore, another key resource of the company is its community of drivers and passengers.

BlaBlaCar develops, maintains and markets a travel-sharing platform that connects drivers with passengers, willing to share the cost of long car journeys between cities. The platform incorporates features such as ratings and reviews of other members and control of social networks. The company also offers a customer service dedicated to its users and manages a network of partners.

The objectives of BlaBlaCar are summarized with the concept of tri-profit: the company's success is assessed in environmental terms (less pollution), in social terms (meeting between different people and travel sharing) and in economic terms.

BlaBlaCar generates revenue by taking a commission on payments made by passengers to drivers. For each payment made by passengers to cover the driver's travel expenses, BlaBlaCar takes 12%. Based on this business model, drivers set a price per seat on which BlaBlaCar adds a small amount to cover 12%.

The company incurs costs in relation to the payment of its 360 employees and the maintenance of the software platform and IT infrastructure. In addition, the company also has costs related to marketing and advertising, as well as fixed costs in the form of rent for its 22 offices.

The BlaBlaCar service is not only innovative from an economic point of view, because the users who use it gain real benefits in terms of savings and flexibility, but also from the social point of view, because the relational dimension that connotes the service is so much a distinctive side of the same, a driver for the users but, above all, an output generated by the same.

BlaBlaCar, redefining the sharing mobility service as a low-brokered model, not only changes the way a company responds to the need for mobility but also fulfils that "re-socializing" function which is one of the biggest promises collaborative economy.

Case 2: Couchsurfing

Casey Fenton, a young graduate, wanted to visit the world, but did not have the slightest intention of staying in a hotel or to move towards a hostel. He, thus, searched the university's online database and sent this request to hundreds of university students. More than 50 young people responded to his appeal after just 24 h.

Choosing his hosts for the period he was staying there, he lived an incredible experience: they did not simply offer him free accommodation, but lent themselves to accompany him to the city, making him visit local places that he would never have discovered. Fenton and his co-founders created a web portal to allow other people to experience similar experiences. It was 2004 and Couchsurfing.com was born.

The name chosen literally indicates the activity of "jumping from one sofa to another" and through its service today the members of the portal choose to share their lives, their experiences with travellers from all over the world, meeting foreigners, the founders are nothing more than "friends who have not yet met". The main objective is to make the trip "a truly social experience".

The social contact is the engine that pushes, on one hand, the hosts to make available a bed, a sofa, any type of accommodation in a totally free, on the other, that moves the surfers to find a place to stop in destinations usually new and unknown.

Intercultural exchange, connection and sharing are the ingredients that characterize the service and that have made it successful.

The people who can get in touch with the Couchsurfing business are mainly university students, adventure seekers and, in general, all those who want to find a free and "social" alternative to traditional hotel accommodation.

What are the needs of the "actors" that Couchsurfing intends to satisfy? In addition to the need for savings, what the company offers is above all membership in a community of travellers, people linked to each other by a sense of sharing, both material goods (sofa and/or bedroom) but, in particular, of different cultures.

The customers to whom the service is addressed are on the one hand the "surfer" travellers who decide to give up the comforts that a hotel can offer and "please" to share a place for the night with other people and on the other the "host", or who provides a room or just a sofa for the "surfer". These are mainly young people, university students and, in general, those who have a strong sense of adventure and adaptation.

Couchsurfing has created a real community of travellers; in this way, all users can relate to each other and compare their experiences.

Couchsurfing not only focuses on providing a free overnight accommodation service but above all on creating a community of travellers, in which the most important values are respect, curiosity towards new cultures and sharing.

To increase the confidence of hosting on the site there are a series of assessments visible to all members, including the profile (with photos and basic information), references (can be positive, negative or neutral) and other useful information for all users.

The main partners of Couchsurfing are the community of travellers, the various hosts present on the territory and all the investors that allow the business to progress.

In addition to the web platform and the entire technological infrastructure, the most important resource of Couchsurfing is its community of travellers, made up of more than 14 million people and 200,000 different cities.

The most important activities carried out by Couchsurfing are those related to the design and maintenance of the web platform, customer support and community management.

Using the service offered by the site, it is possible to reduce all the waste present in the management of a house/apartment, such as water and energy, as well as a significant reduction of emissions into the atmosphere.

Couchsurfing's main goal is to expand the community of people with a great sense of adventure and adaptation, in order to promote a sustainable and social alternative to traditional housing.

The service offered by Couchsurfing is free: guests are invited to offer their help at home to help the host in the various tasks but there is no cash payment. Until 2011, the company defined itself as non-profit and was supported by free donations; in the same year, it was converted into a profit-making company, supported by various investors.

The costs that support Couchsurfing are related to the maintenance of the web platform, the mobile app and advertising.

(Information provided are taken from companies' websites blablacar.com and couchsurfing.com).

3.1.3 Circular Economy

The concept of Redistribution Markets can be seen as being originated by an ideal intersection between Collaborative Consumption and Circular Economy. The idea behind this model is that goods previously owned by someone who no longer needs them can be redistributed to other users. Examples of this redistribution market system can be Freecycle, Share some sugar and eBay. These businesses can consist of completely free trade, sale of goods in exchange for money or credits or even a mix of the two payment methods. Exchanges can be made between complete strangers or between acquaintances belonging to a specific network; the transaction may involve objects of equal value or similar value. The redistribution markets promote the recycling of unused goods in order to reduce waste and extend their life cycle, the term redistribution is part of the 5R list: reduce, recycle, reuse, repair and redistribute. The "greener world" concept and collaborative consumption are certainly fundamental movement for the maintenance of the environment around us (Botsman and Rogers 2010). It can also be said that the exchange of used goods is not really a new phenomenon but rather rooted in past traditions, unfortunately in the twenty-first century this practice is not as widespread as it should be. However, once again, the Internet has revolutionized the concept of exchange and is making redistribution an increasingly attractive move with a significant reduction in transaction costs. Transaction costs are considered as costs for making any form of exchange or participation in a market. Before the advent of the Internet, the transaction costs of unwanted goods were much higher; the correspondence of interests, the management of transactions, negotiations and after-sales activities were not so easy. Now that it is possible to deal in a market without barriers, the application of a redistribution market is practical, beneficial and sustainable.

The redistribution market has two unintended but positive consequences as follows:

- Environmental benefits: The goods are continuously circulating, their life cycle is lengthened, and waste and carbon emissions are reduced. The transport of recycled products has less impact than the production of new items.
- Community construction: The interactions that derive from these systems stimulate the connections between people and consequently the creation of a lasting social capital.

Case Study

The Context of Circular Economy

Case 1: AHLMA

It is estimated that global fabric production is responsible for 20% of industrial wastewater, and emits more greenhouse gases than aviation and international shipping. Above all for this disturbing fact, many companies in the textile sector have implemented the circular model in their supply network, to reduce pollution, waste and improve the efficiency of the company system.

Consider a company in the textile sector, based in Brazil, which is supplied by around 70 dedicated suppliers. It can be classified as a "top-down" company.

Before using the circular model, it used the classic linear model, in which, however, there were many limitations. For example, the large volumes of waste generated can be considered as unexploited opportunities: more than half a billion dollars are lost from the system every year due to underutilized clothing and negligible recycling rates.

Based on these considerations, AHLMA was created to offer a tangible solution through high-quality, economical and unisex clothing lines.

In fact, it has reinvented the operating system of the textile industry, from the conception of the product to the experience of the final customer. The circular model is conceived not only as regards the materials, but is extended to the whole system.

In the traditional model, decisions come down from designers to stylists and then to manufacturers, regardless of the availability of materials. To address this problem, AHLMA brought together industry stakeholders around the idea of decentralized decision-making and collaborative creation of clothing based on current local market conditions, such as considering the actual availability of materials, and adapt to their eventual change.

The activities carried out by AHLMA have been multiple, including the following:

- Over 80% of raw materials derived from the remaining fabric due to the mismanagement of other textile companies. In this way, the use and necessity of raw materials are reduced and, consequently, the production costs of AHLMA's clothes are lower. This entails extra revenue for suppliers.
- "open source" design: All the design and model codes are available on the website, in order to expand the influence on a global level. In this way, they allow anyone to replicate their style to make clothes with specific materials, which are available to them.
- Lean inventory: Since online trade accounts for the majority of sales, with this inventory it is to reduce large surplus cases.
- Use of reusable shipping boxes: In this way, the company wants to encourage the consumer to use these boxes again in other applications.
- Instructions to keep and extend the life of the clothes, available on the website or printed inside the clothes.
- Presence of a physical "concept store", where customers can take advantage of a cleaning service that uses only non-toxic solvents. In addition, in this "store", there is a laboratory to repair and extend the life of clothes, through the maintenance and remodelling of them.

All these activities have produced the expected benefits: positive revenue for producers; in the value chain the relationships between the actors of the supply network have been reinvented, to improve the processes; the fabric represents the lowest cost in company inputs; up to now, over 10,000 items of clothing have been produced using recovered fabrics; there has been a change in the consumer mentality: from fast fashion to conscious fashion.

Case 2: Choisy-le-Roy

The dependency of the automotive industry on raw materials and some precious metals is a major obstacle and presents challenges for supply management. It is estimated that 60% of global supply is destined for car production. In addition to the shortages and difficulties in the procurement of metals, rare or not, the increase in global demand for raw materials has caused costly price increases. For the automotive industry, these additional costs increase by several million euros each year.

It is, therefore, logical that the main concerns of the producers are to be able to anticipate any shortcomings and ensure supply. For this reason, technological solutions have been developed to limit the current dependence on terrestrial metals. According to the Ellen MacArthur Foundation, around 12 million vehicles are collected annually in the European Union, equivalent to millions of tons which are indeed a valuable resource. Using this resource, investing in recycling technologies and increasing the use of recycled material provide a very promising perspective.

The regeneration and the restoration involve the bringing back a part, or a product, to the state as close as possible to the original one, and therefore to its characteristics.

The production of reconstructed car parts began in 1949 in Choisy-le-Roy, located in France. Since then, the factory has constantly diversified its production. In fact, as the years went by, injection pumps were first produced, followed by reducers, injectors and turbo compressors. Nowadays, 325 employees work on the site, producing on demand and guaranteeing the engineering and production of six types of mechanisms.

The regenerated parts are exclusively used for the repair of vehicles currently in use.

The advantages are many, including the cost: this process is 30–50% less expensive, to ensure warranty and safety to their customers, the regenerated parts are subjected to the same quality control tests of the new parts.

By extending the life of vehicles, maintaining value and saving energy, so as to reduce waste, it has created a complete circular model. Furthermore, this activity involves a skilled workforce and creates jobs locally (impact on social dimension of sustainability as well). In fact, in order for this process to be considered economically interesting, the regeneration must be carried out within the market in which the vehicles are used. If the pieces to be processed were shipped abroad to perform these operations, this would not be economically advantageous.

The numbers speak for themselves: 80% less energy, 88% water, 92% chemical products and 70% less waste production.

In terms of raw materials, the plant does not send waste to the landfill. It is estimated that 43% of carcasses is reusable, 48% is recycled in the foundries of the company to produce new parts, while the remaining 9% is valorised in the treatment centres.

This means that the whole process is waste free. As the economy teaches, we must never stop in the development of processes, but we must always try to improve them. As far as the mechanical parts are concerned, they are generally made with a view to repair, but it is possible to go further in the research to improve the materials used and the parts produced. Engineering studies are underway to develop the "future" mechanical parts, for example, by improving their design, in order to make dismantling easier and to increase the recyclability of materials. Other studies are deepening a review of the acceptance criteria and the interchangeability of the components. It is still too early to evaluate the benefits and profits of these different initiatives, but research is going in the right direction by showing it as a very promising market.

(Information and data for the first case AHLMA are taken from the Ellen MacArthur Foundation (ellenmacarthurfoundation.org/case-studies) and from company's website ahlma.cc.

Information and data for the second case Choisy-le-Roy are taken from the Ellen MacArthur Foundation (ellenmacarthurfoundation.org/news/the-circular-economy-applied-to-the-automotive-industry-2)).

3.2 Product Service Systems in B2B and B2C Markets

Servitization requires companies to undergo a considerable transformation according to different elements and characteristics to be taken into account. A key element in this transformation is the destination market or, to better say, the market in which the company operates or the one in which the company wants to expand with the servitized offer. Whether this is going to be a B2B or B2C context, some significant changes occur, and these differences must be taken into account. Table 3.1 contains some examples of servitized offerings in different B2B and B2C contexts.

As it can be seen from Table 3.1, there is a majority of B2B cases emerging from practice: this is mainly due to the fact that PSS started its development in the industrial manufacturing B2B context, where nowadays it has a considerable diffusion. Moreover, PSS-related offerings are more suitable for B2B markets, where providers and customers are more likely to build more rational relationships. For example, Arcelor-

Table 3.1 Examples of different PSSs in different markets and contexts (Laperche and Picard 2013)

Degree of servitization	Categories of PSS	Examples
Low servitization	Customization of manufactured products	B2B • Steel solutions (Arcelor Mittal) • Energy-efficient solutions (STMicroelectronics; Schneider Electric) • Sustainable habitat solutions (Saint-Gobain) • Solutions in energy and construction (Vinci) B2C • Global solutions to reduce environmental footprint (Air Liquide)
	Development of additional product-related services: Information and training	B2B • Training on technology use (STMicroelectronics) • Creation of training centres (Saint-Gobain) • Information and training on energy consumption (Schneider Electric) • Provision on environmental information from eco-design of products (STMicroelectronics, Vinci) B2C • Training on eco-driving (Renault)
	Development of additional product-related services: End-of-life services	B2B • Collection of steel for packaging (ArcelorMittal) • Sorting technologies for recycling centres (ArcelorMittal) • Aircraft dismantling service (EADS) • Collection of vehicles at the end-of-life stages (Renault; JV Indra) • Collection of wastes on construction sites (British Gypsum and Placoplatre; Saint-Gobain) • Collection, recycling, dismantling services (Schneider Electric)
High servitization	Use-oriented services	B2C • Package offering the Z.E. Box (Renault) • Mobility services related to the use of electric car (Renault)

Mittal operates in the steel industry, where it develops and sells solutions linked to lightweight steel for automotive firms; Saint-Gobain realizes insulation solutions (mainly for exteriors) to meet the many needs of companies operating in the building industry; Schneider Electric is focused on the development of energy management systems for a more efficient measurement and management of energy usage. Companies mostly concentrate in two areas for the development of extra services, which are the area of training and take-back/dismantling services. For what concerns training and educational activities, these are extremely valuable in the B2B context, since they allow companies to ensure a better exploitation and utilization of assets, meeting one of the most important drivers behind PSS adoption. Indeed, Saint-Gobain realized training centres in several countries, while Vinci (construction industry) adopts Life Cycle Analysis to allow its customers to monitor the environmental impact of buildings throughout the whole building process. While in some cases, these services have a direct effect on the utilization of assets, allowing for a better usage rate and efficiency in consumption, and in some other cases, these services are aimed at monitoring and providing feedbacks on the physical components of PSSs; in these cases, the effects are more indirect and can be utilized to address environmental issues and legislative pressures.

Other services concern the management of end of life of products and, more in general, of physical components involved in the offering: ArclorMittal developed a system to recollect steel parts used in packaging; EADS manages the dismantling of aircrafts and clean-up (collection) services for military sites; other similar services for take-back have been implemented also by Schneider, Renault (also in B2C markets) and Saint-Gobain, which outsourced this service to other companies (British Gypsum and Placoplatre) in different countries.

On the other hand, in B2C contexts there is a tendency for an "irrational" component since customers are less willing to embrace ownerless consumption (one of the main barriers to PSS analysed in Chap. 1). Indeed, the most successful examples of PSS in B2C markets are those related to sharing economy, and particularly to sharing mobility, i.e. bike and car sharing. It is interesting to note that, even if PSS is more diffused and well accepted in B2B contexts, some of its more radical forms are appearing in the B2C markets, also with a non-negligible rate of development and innovation.

Comparison between different use-oriented PSSs like, for instance, many successful examples of sharing mobility and sharing economy business models shows that this particular offering is experiencing a good answer from market and customers. In the examples provided, the cases present two different offerings (bike sharing and co-working) from different companies with different aims. Indeed, the first one (the advertisement company that developed the bike-sharing model) is a big multinational company with a volume of net sales of more than 700$ million; the second one instead is a small company, operating mainly in Italy and in few other European countries with a limited number of spaces/offerings. Thus, it can be stated that the use-oriented formula, based on concepts of sharing/leasing/renting and pay-per-use payment systems, is probably the case of PSS implementation that faced the most encouraging answer from the market. Instead product-oriented and result-oriented offerings showed non-uniform results in terms of competitiveness, depending on some characteristics that are specific for each offering and business model.

3.3 Product Service System in the Traditional Manufacturing Industries

The traditional manufacturing context is where the PSS witnessed the first and successful experiences linked to its implementation and diffusion. As already evidenced, this is mainly due to the fact that, in this context, the PSS is developed and sold in B2B markets, where companies involved act and take decisions on more rational bases rather than the B2C markets. In this context, we can find the most "simple" examples of PSS development, mainly belonging to the product-oriented category, but we have as well some offerings that involve a more radical degree of servitization. The box below reports the case of a single company developing and offering both product-oriented and result-oriented solutions.

Case Study

Different PSSs, Different Outcomes, Same Company

The first PSS analysed is a product-oriented offering, employed by a firm selling modular carpets and operating in a B2B context: the company sells modular carpet tiles to its customers, including an all-encompassing series of services like installation, maintenance, substitutions, removal and recycle, maintaining a traditional seller–buyer relationship with customers. The company mainly adopted PSS for the development of a circular economy system, mainly related to firm's environmental concern. The interviewed manager (Vice President of Product and Innovation department) clearly recognized the PSS as a source of competitive advantage and extra revenue, thanks to some important services like take-back programs involving dismantlement and recycle of end-of-life products. The most important source of competitive advantage has been recognized as the series of processes involving an entire reconfiguration of the supply chain, in order to implement take-back logistics and recycling facilities in the firm's value chain (hard replication for competitors). Although resources (mainly logistic infrastructures) have been considered quite easy to be replicated, they have been classified as exposed to a medium replication risk, because of the considerable investments required and because of the first mover advantage, of which the analysed company benefitted. Technical capabilities required for PSS design and setup were classified as easy to be replicated.

The same firm offers a second PSS, and both share many characteristics. In this second case, we have a result-oriented offering, where clients lease services of modular carpet system without taking ownership or liability for maintenance: the system comprises the entire series of services presented in the first case (installation, maintenance, substitutions, removal and recycle), including an extra service of design, in order to more specifically meet customer's needs. Although this offering shares with the first one many characteristics and the entire set of resources, capabilities and processes, it proved to be an unsuccessful offering, with a very little number of customers since its launch; according to the manager, this is probably due to the leasing formula that proved to be unattractive for customers, discouraging instead of attracting them.

From the case analysed, it is possible to observe how the introduction of the leasing formula and result-oriented offering brought no significant difference for customers, implying a lack of attractiveness and the inability to constitute an economically sustainable competitive advantage.

The "more simple" product-oriented offering is capable of producing considerable results for the firm differently from the result-oriented one: the main difference here, explaining the evidenced gap in performance, lies in the absence of the leasing formula that is at the core of result-oriented offering. The result-oriented offering, although implemented several years ago, counts nowadays no more than two or three customers worldwide, and this is probably due to the leasing formula that acts as a constraint for customers instead of representing an incentive. Despite both PSSs present a strong environmental concern with the implementation of reuse/remanufacture/recycle services, and this seems to be not enough if compared to the disincentive represented by the leasing formula.

(Information and data presented are taken from interviews conducted by the authors).

The case discussed in exhibit confirms how difficult it can be to generalize out comes related to a specific category of PSS.

PSS was first developed in the manufacturing industries. Table 3.2 reports a list of examples of PSS development in industrial manufacturing companies.

All PSSs offerings needed, for their development, partners' involvement. Indeed, the most frequently required competences concern design of hardware/software as well as approaches like Design for Manufacturing and Design for Assembly; together with these, the involvement of suppliers at various levels (component and subsystem suppliers), and customers' specifications. Another evidence is that the development of integrated products is not enough as a competence to fully deliver a PSS-related offering to the market. The majority of cases proved that there is a tangible need for business networks, as already highlighted. Furthermore, to be successful, a company undergoing servitization needs to manage and be involved in multiple business network realities at the same time. This is mostly true in case of SMEs, which are companies that usually can benefit of a considerable specialization, but are in need of business partners in order to be competitive and succeed on the market.

Table 3.2 From traditional products to PSS for manufacturing firms (Isaksson et al. 2009)

Company	Traditional product	PSS	Key competences required
Toyota Material Handling Group	Forklift trucks	Warehouse transportation solution	• Customers' internal logistics • Remanufacturing
Volvo Aero	Jet engines	Providing thrust on the wing "power by the hour"	• Flight operations • Airline customer's strategy
Marine Jet Power	Water jets	Water jet propulsion systems	• Customers' jet board operations
Sandvik Coromant	Tools for metal working	Smart manufacturing	• Customer machine investments • Customer value analysis
ITT Flygt	Pumps	Cleaning solutions	• Functionality of customer plants for wastewater
Rodeco	Plastic figures	Adventure pool parks Playgrounds	• Delivery, assembly and functional testing • Playground competences • Children's safety standards • Facilities management
Svenska Expander	Spare parts for machines	Expander system ™	• Customer productivity • E-business system for mass customization
Polyamp	DC/DC converters AC/DC power supply	Naval systems for mine countermeasure applications	• Customer training • Reduced ownership costs • Naval industry use • Software package for customer use

(continued)

Table 3.2 (continued)

Company	Traditional product	PSS	Key competences required
Assalub	Lubrication equipment components	Lubrication systems	• Knowledge about customer needs for reduced consumption
Storebro maskinrenovering	Contract manufacturer	Machine reconditioning and preventive maintenance	• Systematic de-assembly and reassembly • Problem-solving • Collaboration with programming firms • Operator training
Ocean Modules	Underwater robots	Underwater operations	• Customers' underwater operations
Industrihydraulik	Contract manufacturer	Hydraulic systems	• Hydraulic systems for customer activities

Furthermore, the modularity of services and products is often indicated as a key requirement for the successful development and delivery of a solution in the manufacturing context. The necessary interrelations between products and services and the exclusiveness of many of them might represent a relevant obstacle to PSS implementation and/or to the implementation of different solutions inside the same company. Adopting a modular approach can allow more favourable conditions to PSS spread and adoption among manufacturing companies.

3.4 Product Service System in the Sustainability-Driven Industries

The concept of servitization and PSS, since its origins, has always been closely related to the topic of sustainability, for its ability to impact all the three dimensions of social, environmental and economic sustainability.

Some of the most recurring examples refer to PSS solutions developed in the field of the so-called sustainable industries, i.e. those sectors where sustainability is one

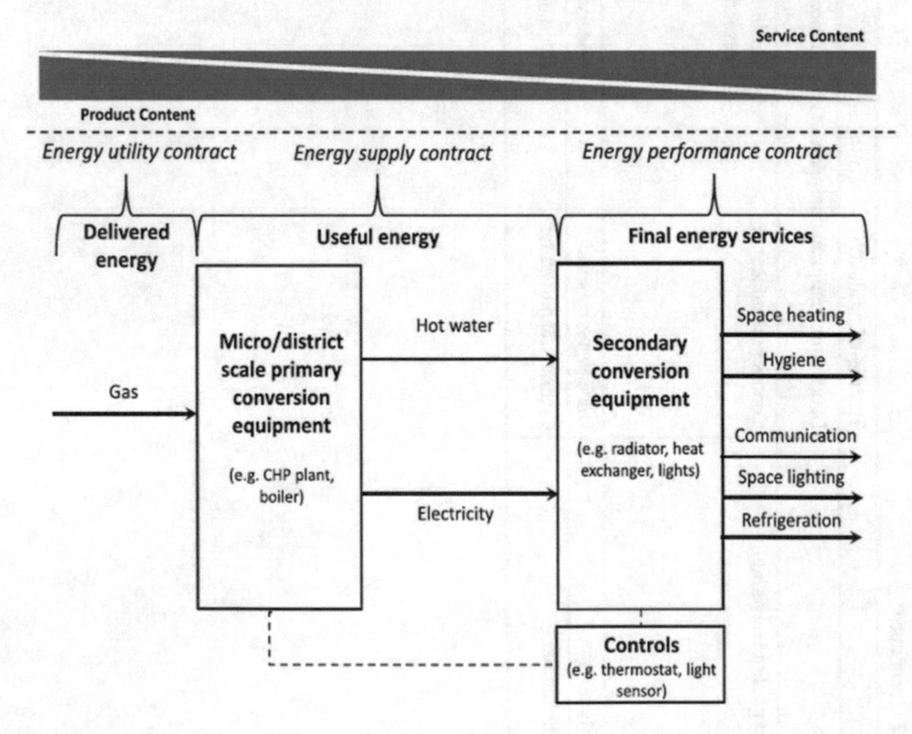

Fig. 3.4 PSS-derived contracts concerning energy utilities and energy services companies (Hannon et al. 2015)

of the major goals of companies. For instance, the energy sector can provide many insightful examples of PSS development. Figure 3.4 shows a framework depicting different types of PSS-deriving contracts in the energy service field.

The framework reports three different types of contracts located along the product–service continuum: indeed, the energy service companies context can be considered as an energy subset of PSS. Customers subscribe energy service contracts for the provision of energy services, constituting “the transfer of decision rights over key items of energy equipment under the terms and conditions of a long-term contract, including incentives to maintain and improve equipment performance over time” (Sorrell 2005). The basic model on the left of the framework represents the “normal” situation where users pay for energy provision. On the other hand, energy service contracts can be distinguished into two different categories that are “Energy supply contracts” and “Energy performance contracts”. The first kind of contract resembles a use-oriented offering, where the company provides customers with energy streams, and payment is based on “unit per useful energy” (Sorrell 2007) or a fixed price according to a predetermined level of supply (Marino et al. 2011). The second type of contract can be considered as a result-oriented model: customers are, in fact, provided with a “final energy service” (Hannon et al. 2015). The framework also indicates which parts of equipment are maintained under the energy service companies ownership and control throughout the whole duration of the contract.

Another interesting example concerns the development of PSS in the field of solar home systems (Friebe et al. 2013). Figure 3.5 contains four examples of PSS: the first two are related to a sale model (basic PSS offerings, similar to the product-oriented category), while the second two refer to a service model, with a leasing and result-oriented offering.

In the cash model, the customer pays for the product (solar home system) which is installed by the provider or by the customer itself: there is, in this case, a shift in ownership and a minimal service component in the offering. The model is so called since, given the nature of the transaction, cash payment is (expected to be) the preferred form of payment.

The credit model is very similar to the basic cash model, but it includes a financing service (a loan) that can be provided from a financial institution or by the company itself.

With the leasing model, the offer becomes more servitized: the customer pays for the use of the solar home system and the ownership is maintained by the company, and there can be a shift of product ownership to the customer if the solar home system is fully payed, like for every leasing contract. This model usually involves a financial partner, and since the company maintains the ownership for a considerable amount of time, an advanced service component (e.g. sales service and maintenance) is provided.

The last model, named Free-for-service, is almost completely a result-oriented offering, where the company maintains the ownership of the solar home system, and the customer pays for its usage on the basis of a regular fee, or according to the consumption (calculated, for instance, in kWh used). Obviously, in this model, the company provides a full-service offering, involving first of all preventive maintenance

	Sales model		Service model	
	Cash	Credit	Leasing	Fee-for-Service
Market potential	Low (<3%)	Medium (<20%)	Large (<50%)	Large (<70%)
Ownership	Consumer becomes owner upon payment	Consumer becomes owner through contractual agreement	Service provider is owner during the leasing period, then consumer	Service provider
Initial investment cleared by	Consumer	Financial institution plus down payment by consumer	Service provider and eventually Financial institution	Service provider
Regular instalments	No	Yes, to cover the credit	Yes, to cover the rent	Yes, to cover the use of service
Responsibility for maintenance	Consumer	Consumer and eventually service provider	Consumer or Service provider	Service provider
Typical maintenance service	No	Often included for a certain time period	At least included during payment period	Included during contract duration
Major risk for consumer	High technical risk	Low technical risk	Low technical risk	Very low risk
Major risk for service provider	Technical risk covered by manufacturer, low financial risk	Technical risk and eventually financial risk	High technical and financial risk	Very high technical and financial risk
Major risk for financial institution	n.a.	High financial risk	Medium financial risk (for refinancing the service provider)	Medium financial risk (for refinancing the service provider)

Fig. 3.5 PSS models for solar home systems (Friebe et al. 2013)

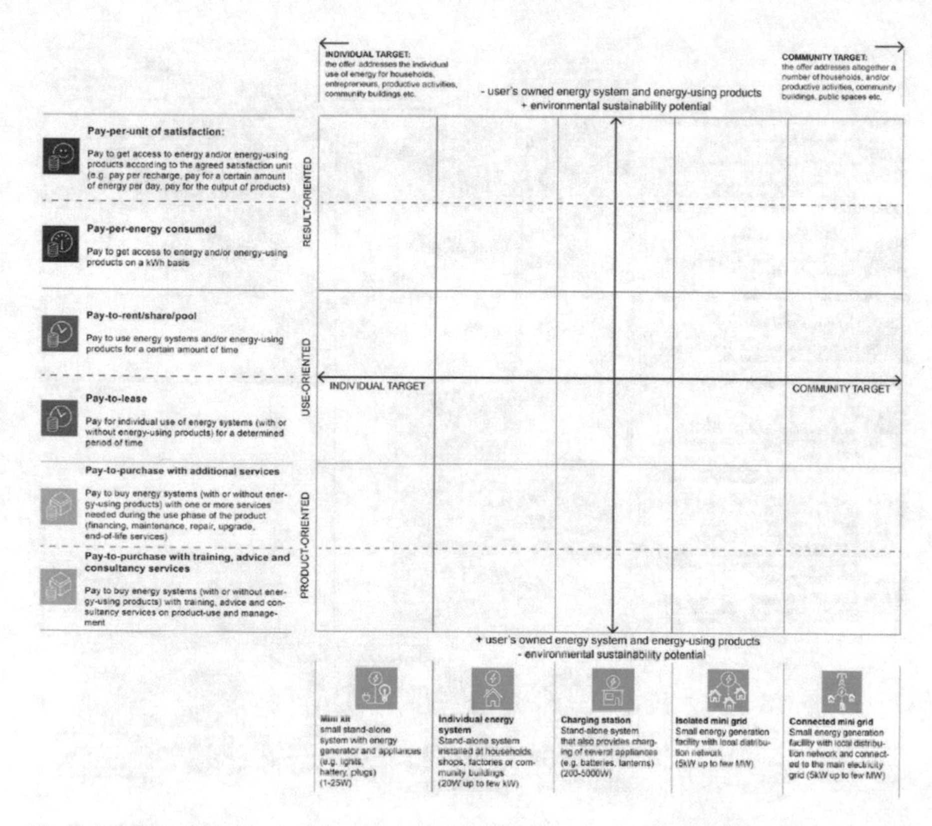

Fig. 3.6 PSS classification in the sector of renewable energies (Emili et al. 2016)

and performance monitoring, to ensure the correct functioning of the system and a satisfactory level of service. Obviously, this PSS mostly involves a non-negligible amount of risk for the producer/provider.

The field of renewable energy or, to better say, sustainable solutions presents a considerable variety of offerings and business models, as can be evinced by the examples proposed. Figure 3.6 shows a classification framework for different PSS cases in the field of renewable energies, developed by Emili et al. (2016).

The vertical axis reports PSS categories divided in product, use and result oriented: as shown in the graphic, the higher we go along the axis, the higher is the environmental sustainability potential of the PSS considered. Indeed, in the lower extreme, we have two basic offering models, where the product (energy system) is sold in the first case with basic services (training, advice and consultancy) and in the second one with additional services like, for instance, financing, maintenance and repair. In these cases, the gain in sustainable performance minimally differs from the pure product offering. The central part of the axis, focused on the use-oriented category, contains two models based, respectively, on a lease formula and a renting/sharing formula. These modes of offering presents non-negligible environmental benefits,

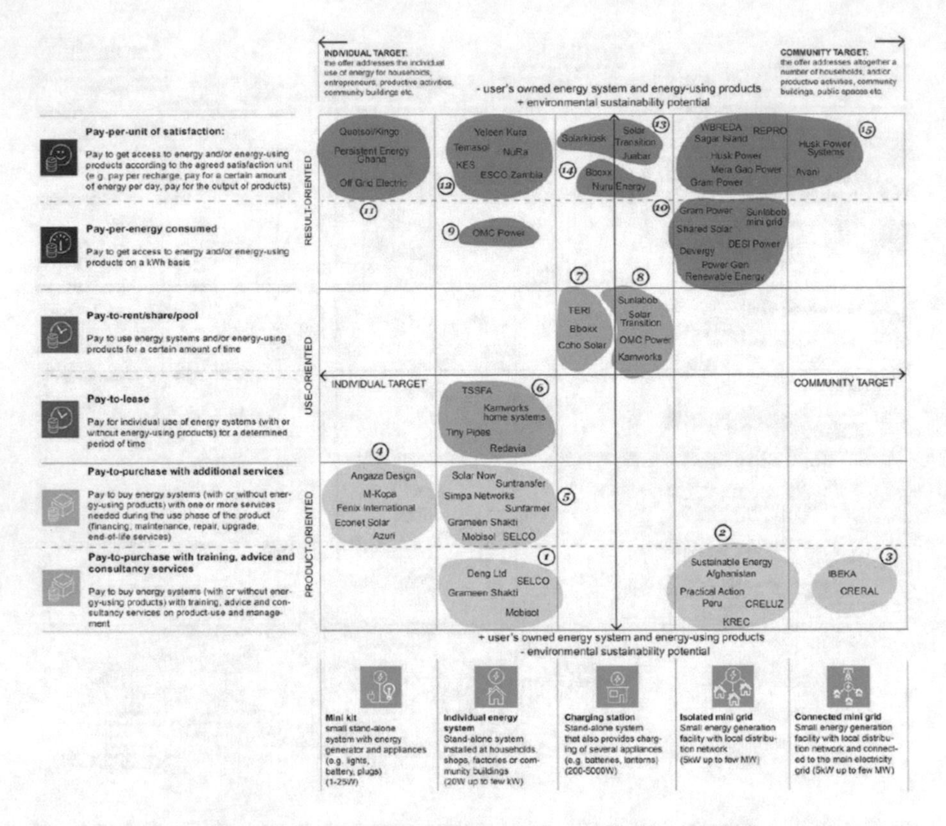

Fig. 3.7 Archetypal models in the PSS classification scheme (Emili et al. 2016)

since they ensure a prolonged life cycle for physical components involved in the offering, and this imply a reduced rate of production with a consequent reduced amount of inputs and wastes. Finally, the higher extreme of the axis presents two result-oriented models, named "Pay-per-energy consumption" and "Pay-per-unit of satisfaction": in these cases, we have the highest possible degree of servitization, with the most relevant impact in terms of environmental sustainability.

The horizontal axis, on the other hand, represents the target, ranging from "individual" (left side) to "community" (right side). In the lower part of the framework are also reported examples of offerings covering this range: there are, for instance, "Mini kit" for individual domestic use, and "Grids" which are energy facilities with the capacity of serving a wide range of users.

Figure 3.7 shows the same framework with some "archetypal models" represented in it.

These 15 models are as follows:

1. *Selling individual energy systems with advice and training services:* this is a basic product-oriented model (Fig. 3.8).

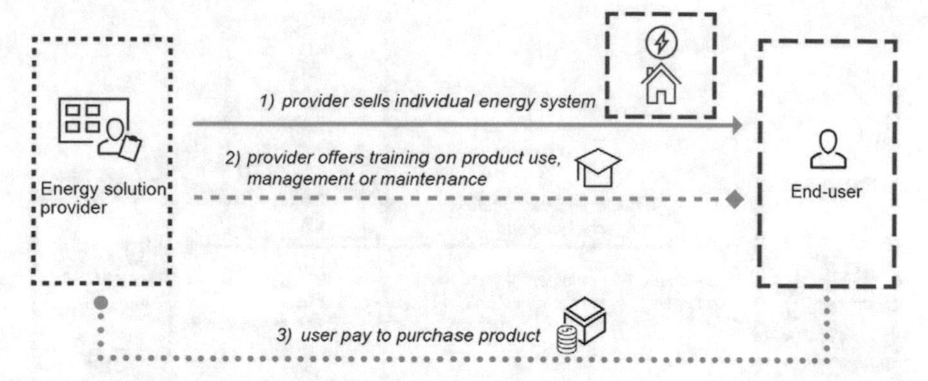

Fig. 3.8 Archetypal model 1: selling individual energy systems with advice and training services (Emili et al. 2016)

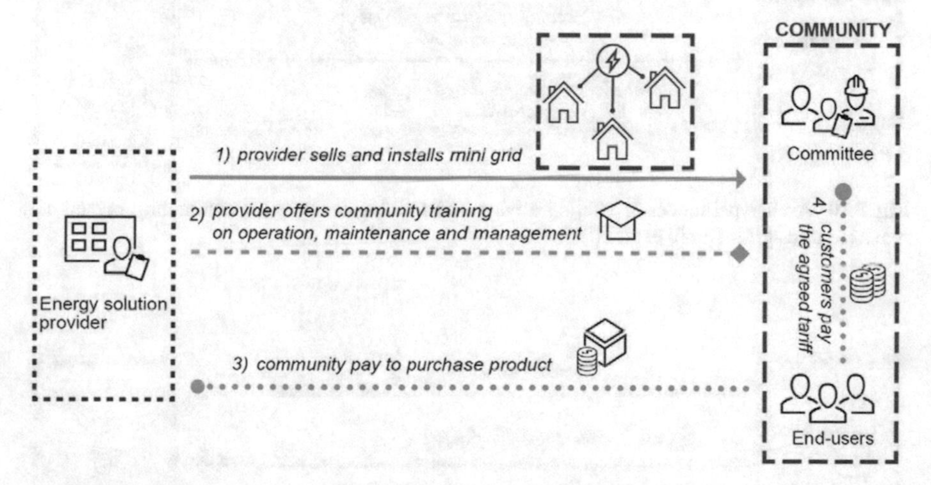

Fig. 3.9 Archetypal model 2: offering advice and training services for community-owned and managed isolated mini-grids (Emili et al. 2016)

2. *Offering advice and training services for community-owned and managed isolated mini-grids:* In this case, the customer is not an individual but a community, even if the nature of the offer itself does not actually change (Fig. 3.9).
3. *Offering advice and training service for community-owned and managed connected mini-grids:* This is very similar to the previous model, with the only difference that PSS sold is connected to a local or national network of energy service, rather than being isolated as in the previous model (Fig. 3.10).
4. *Selling mini-kits with additional services:* These extra services include, for instance, financing, or some basic forms of maintenance (which, however, is mostly up to the customer) and training/consultancy (Fig. 3.11).
5. *Selling individual energy systems with additional services:* In this case, the difference with the previous model concerns the physical product offered, which

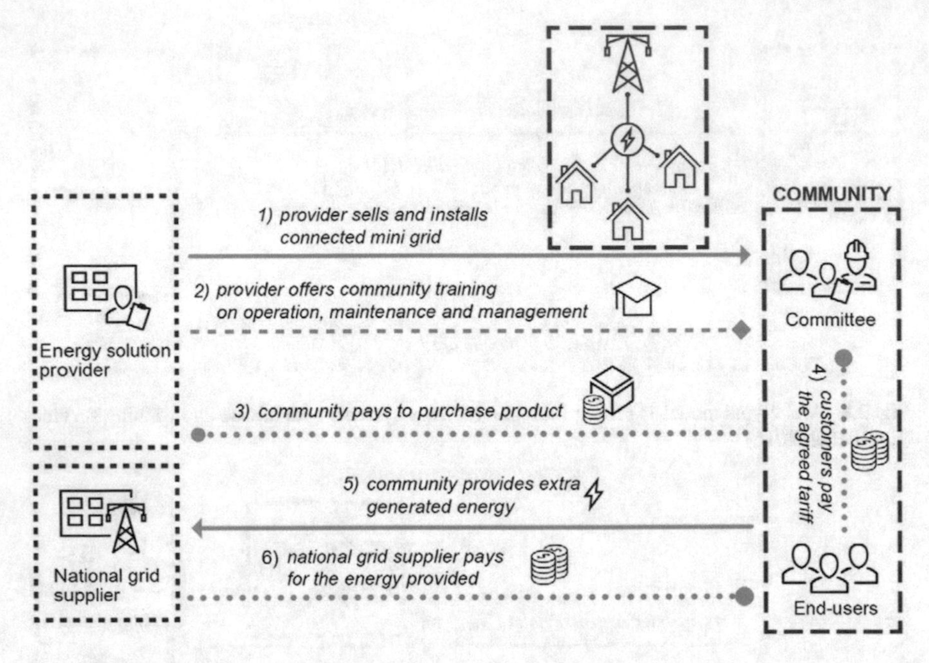

Fig. 3.10 Archetypal model 3: offering advice and training services for community-owned and managed mini-grids (Emili et al. 2016)

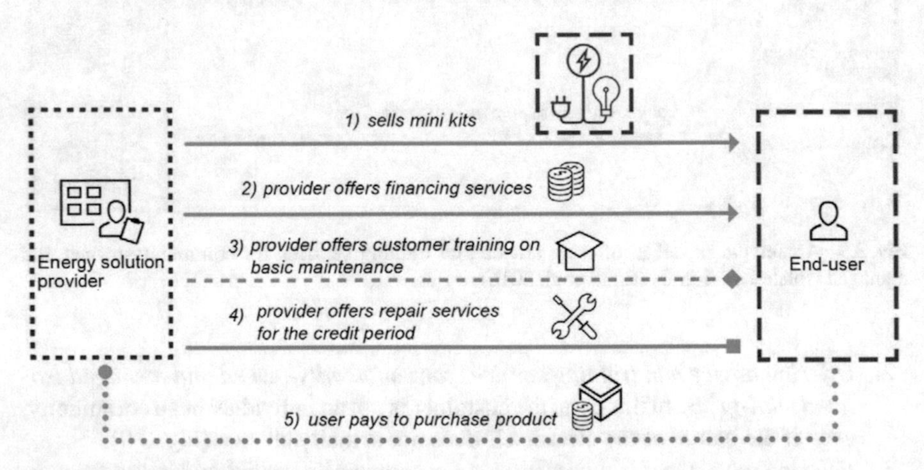

Fig. 3.11 Archetypal model 4: selling mini-kits with additional services (Emili et al. 2016)

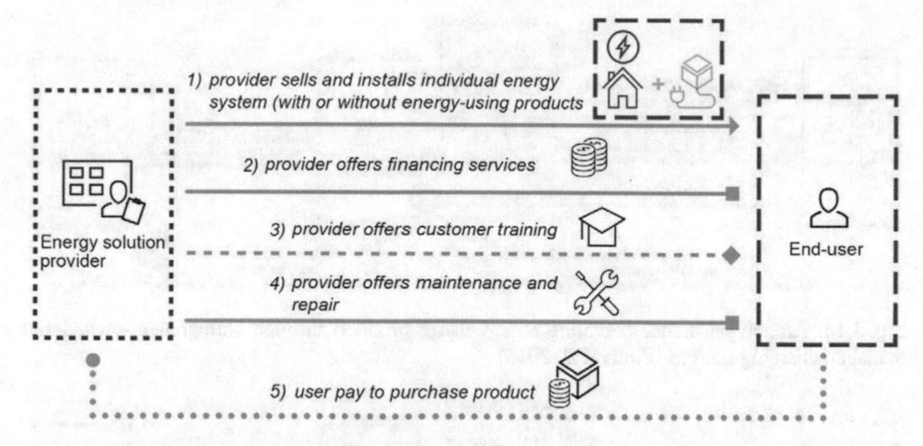

Fig. 3.12 Archetypal model 5: selling individual energy systems with additional services (Emili et al. 2016)

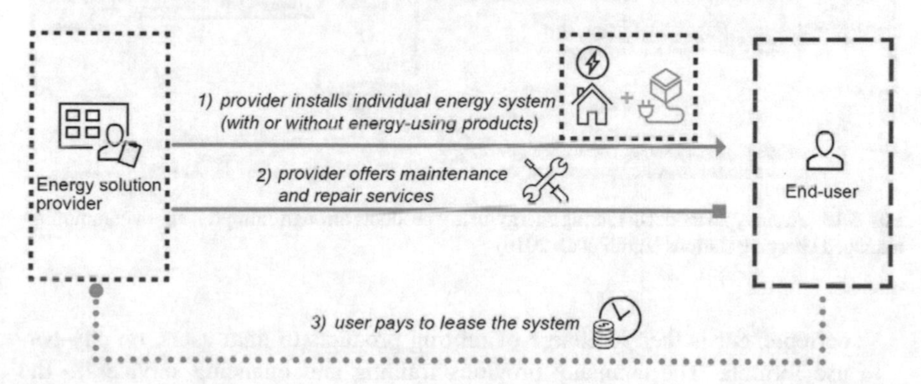

Fig. 3.13 Archetypal model 6: offering individual energy systems (and energy-using products) in leasing (Emili et al. 2016)

is not a mini-kit but an individual energy system. In this case, the more complex nature of product sold implies some differences in additional services offered to customer, for example, more specific installation services and/or training, maintenance and repair services (Fig. 3.12).

6. *Offering individual energy systems (and energy-using products) in leasing:* This is the first model which does not involve a direct and immediate shift in ownership of products. Indeed, as the name itself suggests, the model adopts a payment formula based on leasing, and the company provides a whole set of services like repairs and maintenance, to ensure the product's durability (Fig. 3.13).
7. *Renting energy-using products through entrepreneur-owned and managed charging stations:* In this case, the charging station is sold together with energy-using products to a local entrepreneur, with a shift in ownership. The

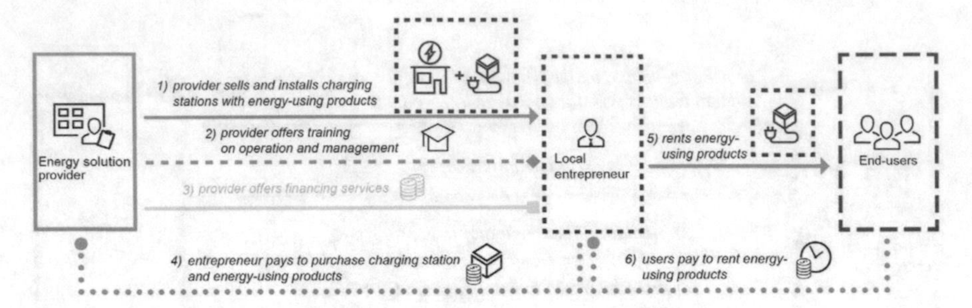

Fig. 3.14 Archetypal model 7: renting energy-using products through entrepreneur-owned and managed charging stations (Emili et al. 2016)

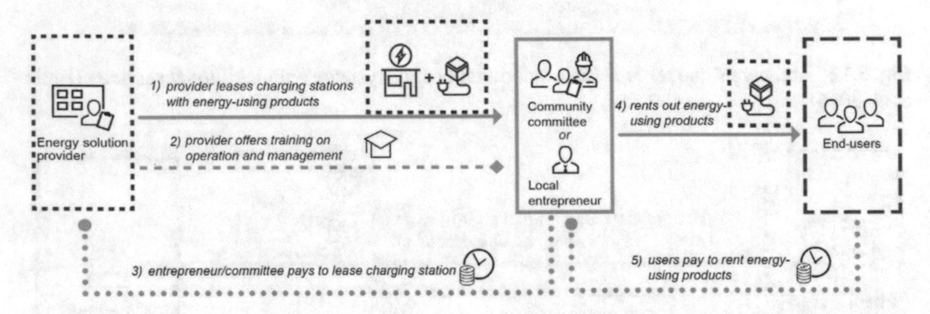

Fig. 3.15 Archetypal model 8: renting energy-using products through entrepreneur- or community-managed charging stations (Emili et al. 2016)

entrepreneur is then in charge of renting products to final users, on pay-per-use formula. The company provides training and financing services to the entrepreneur, who is then responsible for maintenance in use of both charging station and products (Fig. 3.14).

8. *Renting energy-using products through entrepreneur- or community-managed charging stations:* The company installs a charging station that will be used by final customers with energy-using products. In this case, the company maintains the ownership of the charging station but its management is in charge of a local entrepreneur or of the community of customers/users, who pay a fee for using the station and a rent for the products (Fig. 3.15).
9. *Offering access to energy (and energy-using products) on a pay-per-consumption basis through individual energy systems:* In this case, there is an individual energy system installed for energy need of the customer, which pays according to the energy consumption. The company maintains the ownership of the system and provides all necessary service to keep it in function and ensure its durability (Fig. 3.16).
10. *Offering access to energy (and energy-using products) on a pay-per-consumption basis through isolated mini-grids:* Mini-grids at a community level are installed by the provider, which maintains the ownership, and users

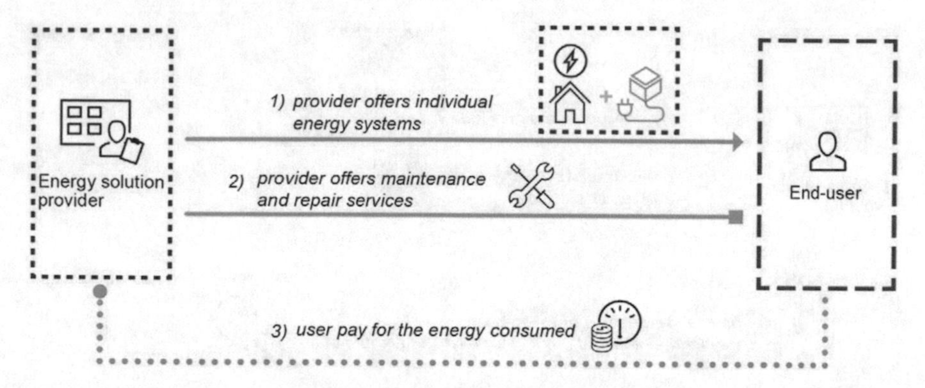

Fig. 3.16 Archetypal model 9: offering access to energy (and energy-using products) on a pay-per-consumption basis through individual energy systems (Emili et al. 2016)

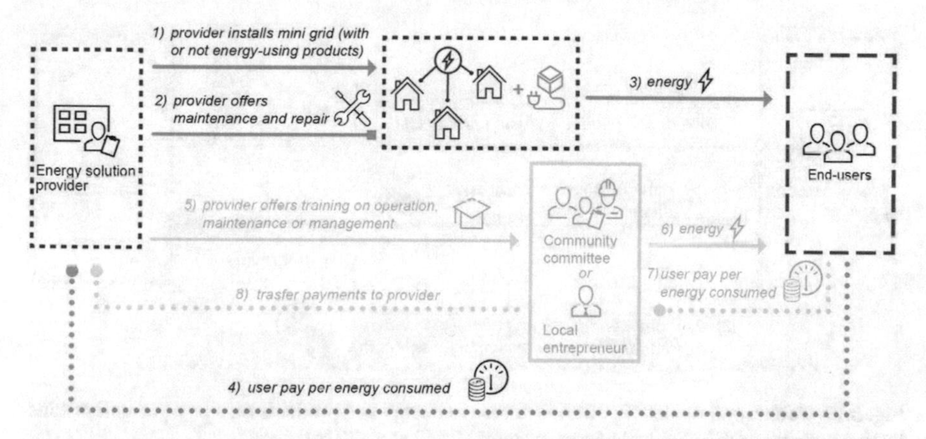

Fig. 3.17 Archetypal model 10: offering access to energy (and energy-using products) on a pay-per-consumption basis through isolated mini-grids (Emili et al. 2016)

pay according to the usage of the system. In this case, a local entrepreneurship of the community itself might be in charge of managing and maintaining the mini-grids. If the entrepreneur is involved, it is in charge of collecting consumption-based fees and transferring them to the provider (Fig. 3.17).

11. *Offering access to energy and products on a pay-per-unit of satisfaction basis through mini-kits:* The provider installs mini-kits with energy-using products, and customers pay according to a service package chosen among those offered by the provider. The overall offering includes always maintenance, repair and other services since the provider maintains the ownership of all physical components/products involved (Fig. 3.18).
12. *Offering access to energy and products on a pay-per-unit of satisfaction basis through individual energy systems:* The model is very similar to the previous one, except that users pay for individual energy systems on a monthly fee in

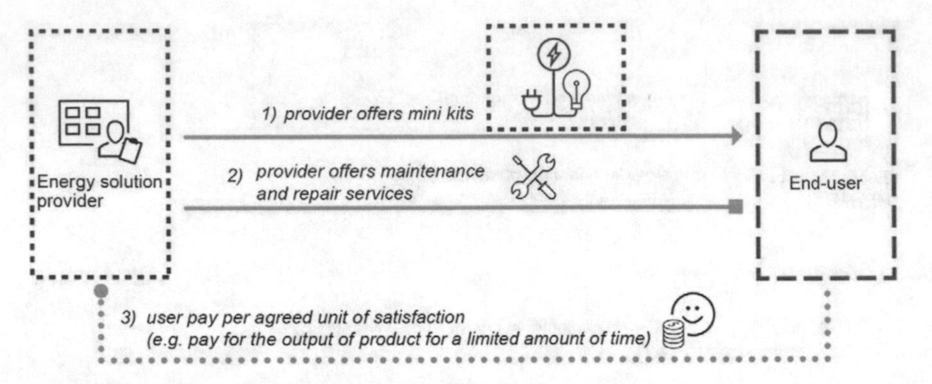

Fig. 3.18 Archetypal model 11: offering access to energy and products on a pay-per-unit of satisfaction basis through mini-kits (Emili et al. 2016)

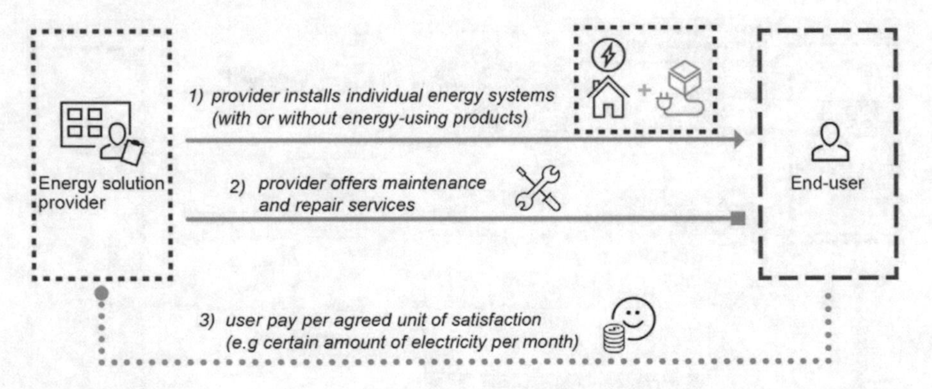

Fig. 3.19 Archetypal model 12: offering access to energy and products on a pay-per-unit of satisfaction basis through individual energy systems (Emili et al. 2016)

order to have access to energy/electricity for a fixed amount of hours each day (Fig. 3.19).

13. *Offering access to energy-using products through community- or entrepreneur-managed charging stations on a pay-per-unit of satisfaction basis:* This model resembles the offering of model 8, but, in this case, the entrepreneur or the community is in charge of providing a set of energy-related services (from printing to IT service, for instance), and users pay on a pay-per-unit of satisfaction basis (like the number of copies printed). The entrepreneur/community is in charge of managing the system, operating and maintaining it, and transfers part of the profit to the provider (Fig. 3.20).
14. *Offering recharging services through entrepreneur-owned and managed charging stations:* in this case, there is only a local entrepreneur involved and it offers only recharging service provided by charging stations. The entrepreneur owns the system and is responsible for all related services, and like the previous model transfers part of profits to the provider (Fig. 3.21).

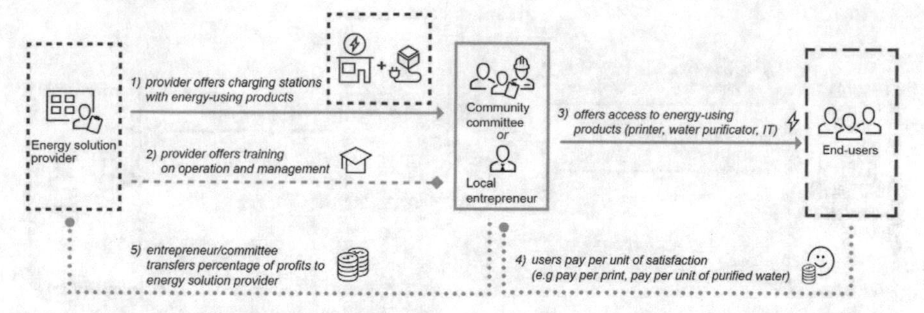

Fig. 3.20 Archetypal model 13: offering access to energy-using products through community- or entrepreneur-managed charging stations on a pay-per-unit of satisfaction basis (Emili et al. 2016)

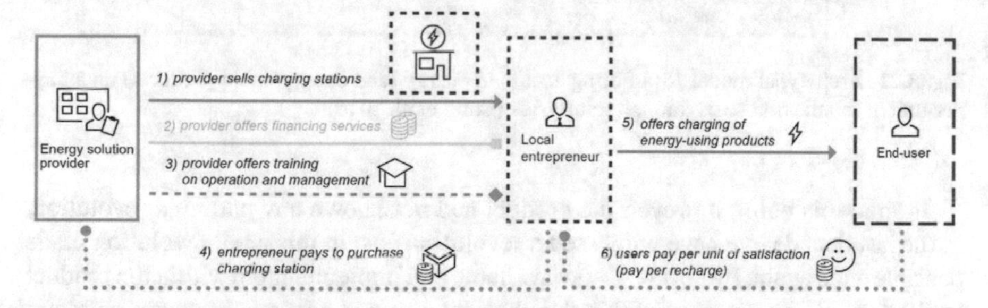

Fig. 3.21 Archetypal model 14: offering recharging services through entrepreneur-owned and managed charging stations (Emili et al. 2016)

15. *Offering access to energy (and energy-using products) on a pay-per-unit of satisfaction basis through mini-grids:* The provider in this case installs mini-grids (that can be connected or not to the main grid) and related products, and users pay for access to energy for a determined amount of hours a day (Fig. 3.22).

3.5 Product Service System in the New Digital-Driven Industries

The spread of servitization has triggered within the organizations a series of important structural and procedural changes, which are often influenced by the presence of a parallel path of digitalization, i.e. the increasing use of digital technologies within companies that allow them to connect with each other. People, systems, companies, products and services are enabling new and incredible business opportunities.

Surely, over the past 50 years, ICT has led to an incredible series of innovations in the business and competitive environment, and it is no coincidence that the responsible for the so-called Third Industrial Revolution took place in the second half of the twentieth century.

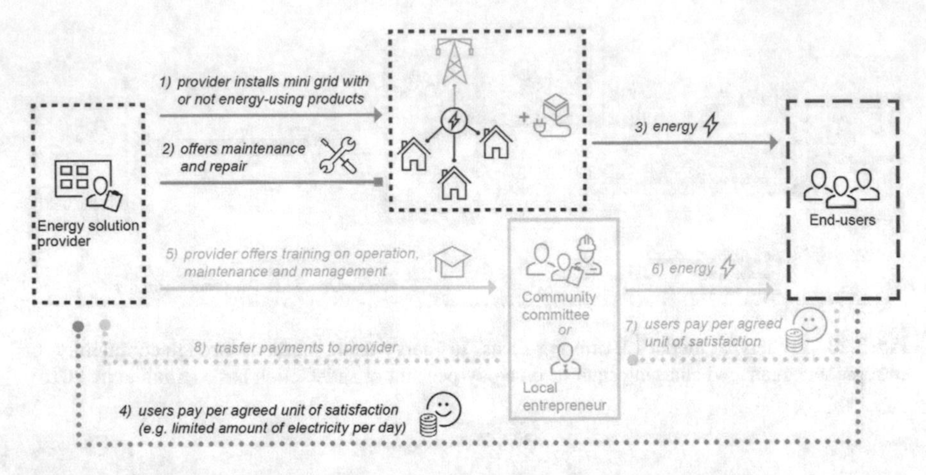

Fig. 3.22 Archetypal model 15: offering access to energy (and energy-using products) on a pay-per-unit of satisfaction basis through mini-grids (Emili et al. 2016)

If up to this point, however, the product had not known any particular evolution, in the last decade, we have witnessed a revolution also in this area, revolution made possible once again, by new ICT now available and implementable within the product itself: through the presence within the physical assets of advanced sensors, software and integrated processors, in fact, it has now become a real window on the customer, which generates data in real time during use and shares it with the supplier through a direct connection with a persistent online platform. This fact enables a whole series of advantages, both for the customer in terms of functionality and performance, and for the supplier, who at this point has a huge amount of usage data that can be exploited both to improve the service offered to its current customers, both to achieve targeted and effective future innovations. Supporting the production and delivery of this new type of product, however, is not simple and requires an important restructuring of the entire value chain and business processes, from design to post-sales, so much that in recent years to define the phenomenon and its impact Fourth Industrial Revolution term has been used.

To make this evolution of the global economy feasible there has been a series of technological advances and contextual competitions, such as the increasing miniaturization of electronic components, the improvement of energy efficiency, the availability of high computational power and low-cost storage capacity, combined with the development of tools that now allow faster software development and reduced investment (Porter and Heppelmann 2014). In support of this, we must not forget the release of the new version of the IPv6 Internet communication protocol, which has now made available the huge amount of 340 trillion new Internet addresses, able to be associated with every single "smart" device.

In order to fully exploit the full potential of this new type of product, the development of a real technological infrastructure is required, which can be achieved through major investments and involves various aspects of the organizational situation.

Central element of connection is the online structure of the company: within the cloud there are databases useful to store data from the installed base and tools that allow to manage, standardize and above all interpret them, so as to extrapolate customer information and be able to plan adequate actions to respond efficiently and effectively to any emerging problems or needs. All company functions are also connected to it, as the main business systems (e.g. CRM, ERP) are integrated into the digital ecosystem and can draw on the databases of products using them to optimize their activities. In such a connected reality, with so many sensitive data potentially accessible from any location, it is not surprising that cybersecurity now plays a critical role: any company that decides to embark on a digitalization path will therefore have to invest considerable resources in this area, to guarantee the protection of information on the platform from external aggressors and avoid losing the substantial competitive advantage present in them.

Considering the complex nature of digitalization, it opens up a whole new set of possibilities to companies, but at the same time it carriers within itself a series of barriers to implementation.

A first barrier to overcome is the resistance to change of their own employees (Iansiti and Lakhani 2014): condition inherent in the human soul, must be approached carefully because there is nothing more difficult that convincing a person to change their beliefs or habits, and implementing digital technologies within their organization implies major changes. In fact, to fully exploit the potential of the digital revolution, the various company functions must open their own "information silos" and understand the importance of cooperation with other offices and data sharing. A concrete example to make this change of mentality less traumatic comes from General Electric (Iansiti and Lakhani 2014), which at the time of uploading online its shared technology platform has given its divisions the freedom of choice. In a short time, the improvements obtained in terms of revenues and performance became evident to the entire organization, and even the most reluctant business units were convinced of the goodness of innovation, also entering the IT platform.

Going instead to examine the most concrete obstacles that could prevent a company to opt for a digitalization of its products, certainly a prominent position occupies the need for a communication network developed and widely present, which connects the entire organization through its technological infrastructure and allows the management of large amounts of data (Lerch and Gotsch 2015); currently only the most structured companies can guarantee a similar amount of investments, but fortunately this condition is a condition sine qua non only for the provision of advanced digital services, while in order to offer services of lower complexity many organizations can adopt a more gradual path, which may mean the digitalization of only a portion of its value chain (Coreynen et al. 2018). Regardless of the path of evolution that a company wants to follow, it cannot be ignored, however, that implementing digital elements within your organization requires a large amount of liquids, and that the results of these investments can be seen only over time, when the digital ecosystem begins to take shape. For many companies, a "leap into the void" of these proportions may not be easy to accept.

Less difficult economically but not less important is the difficulty to find qualified employees who are able to develop and manage the innovations that digitalization implies for the company processes and for its offer (Lerch and Gotsch 2015). As also underlined in the previous chapter, the ideal profiles for units in direct contact with the user now have interdisciplinary skills, which make them able to face the increased complexity of a digital offer, interpreting the amount of data of smart products and supporting the customer in the extrapolation of information from them. Personnel of this kind is currently rare because only from a few years training paths specific to this type of role have been created, and therefore organizations are forced to hardly contend for the few available profiles. The start of a digitalization process, however, could also imply a further problem in this respect: not all the companies that evaluate to digitize themselves have since the beginning an IT function; organizations that find themselves in this situation could see as a major obstacle the necessity of having to enter into a territory for them unexplored, creating and structuring a new function within the organization chart that guarantees the resources with the appropriate technical skills to develop and then manage its own technological infrastructure (Porter and Heppelmann 2015).

Last element of interest on the subject, the amount of complexity that digitalization brings with it, however, can never be faced by organizations in an independent manner, and requires important inputs within its value chain, which must be carefully evaluated: more specifically, in order to build their own digital ecosystem, many organizations have established links of various kinds with important players in the electronics and IT industries, sometimes even competitors, creating joint ventures, partnerships and sometimes investing in crowdsourcing projects to distribute risks and responsibilities (Iansiti and Lakhani 2014). Moreover, given the strong electronic component present in the smart products needed to enable the various digital capabilities, to develop a digital proposal, it is essential to accept the entry into its own supply network of electronic component manufacturers, which very often are giants of the world economy (e.g. Intel, Google): there is therefore a real risk that these can catalyse the value supplied to the customer on them, thus reducing the profits of the producer as well as the power in the supply chain (Porter and Heppelmann 2014).

Other than barriers, drivers of digitalization must be taken into account as well. The most important of these factors is probably the willingness of organizations to improve their knowledge of customers (Porter and Heppelmann 2015): as already underlined many times in this chapter, in fact, the main contribution of digitalization to the corporate cause is the incredible production of data that it allows, data coming directly from the operations of its users, and therefore loads of potential information that if exploited opportunely can generate value and new business opportunities; by analysing them appropriately and crossing them with other data, it is now possible to arrive at a deep and detailed knowledge of the client's reality, understanding what his needs are and even identifying any hidden needs. As can be understood, very often these business opportunities are captured not by modifying the physical product, but by updating the service component of the PSS.

This last observation introduces to two other important drivers of digitalization, which concerns precisely the possibilities that it implies for the offer. One of the most important drivers of servitization is its ability to extend the organization's proposal, making it at the same time more dynamic and able to adapt to the needs of the case; developing digital capabilities, the potential of this choice can be fully exploited: they, in fact, make more advanced and complex services possible (Lerch and Gotsch 2015), while allowing greater customization of the PSS at a lower cost and with faster timescales (Porter and Heppelmann 2015), concretely enough to think about the possibility of remotely updating the software present in the products, improving the performance and the service provided to the user without this involving long periods of maintenance or assistance. Customization with digitalization is also now available and easy to handle not only for PSS services modules, but, thanks to the spread of instrumentation for "additive manufacturing" such as 3D printers, also for product modules, guaranteeing to the front-office units that are closer to the final user the ability to make small changes to the standardized proposal assembled initially and provide it in a short time to the customer.

Finally, it cannot be neglected that developing a digital infrastructure also contributes to raising the entry barriers in its market (Porter and Heppelmann 2014), when a provider manages to develop a complex and integrated ecosystem like the one shown in Sect. 3.1, for a potential competitor who does not have the same facilities becomes very difficult to achieve adequate competition. Given the amount of resources that requires its construction, there are few companies that cannot even afford to evaluate a possibility of this kind; moreover, these would then be confronted with a user closely linked to the other organizations already present in the market through the smart products and services offered to them, making the situation even more difficult.

By critically analysing the digitalization phenomenon, it appears how it is interconnected with servitization, for what concerns the respective barriers, drivers and contextual factors.

The links between the barriers of the two phenomena are multiple: the need for a network of facilities distributed near the customer base, for example, which until sometime ago was one of the major obstacles to the implementation of services within its own offer, is now considerably relaxed by digital transformation, as it guarantees the tools to deliver remote monitoring and control proposals (Kindström and Kowalkowski 2014).

Nevertheless, even the much-feared "service paradox", i.e. the fall in total profits despite an increase in turnover that can follow the implementation of a service-focused offer (Gebauer et al. 2005) can be considerably mitigated by digitalization. In fact, the transition to a PSS policy can be made less traumatic and more scalable, giving the possibility in case of lack of resources or need to be able to choose whether to "serve" or before the back area office, thus improving the efficiency of its internal processes, or that of the front office, thus ensuring a more intense relationship with its customers and greater knowledge of their needs (Coreynen et al. 2018). Other two barriers, i.e. the lack of personnel with the appropriate skills to support the evolutionary process for organization and resistance to change are common to both phenomena: the required skills are different for the two phenomena, but what matters is that for both the profiles sought are at this time difficult to find, and this can discourage companies to undertake the path. Also, the resistance to change has different nuances: in the case of servitization, it concerns the change of mentality that implies accepting the risks of managing a product for the whole life cycle and considering the physical product no longer as the heart of the offer (Coreynen et al. 2018), while digitalization of the concept has different connotations as the latter changes the internal company processes themselves, requiring the various functions to cooperate closely with each other and to open their "information silos" to the others, which in the organizational context is not easy to accept.

As regards drivers, the two phenomena have many influences. First of all, there is a clear link between the first drivers of the two categories: if, in fact, many organizations decide to develop a service-based proposal to be able to establish long-term relationships with their clients, this is certainly facilitated by the deep knowledge of their processes, made possible by the data obtained from its installed base when it includes digital instruments such as sensors able to capture them, and software enabled to interpret and transform them into information. Even the long-term criticalities linked to ethical and environmental aspects, thanks to the characteristics of smart products, may be improved. Thanks to the presence of digital hardware, many product features can be simply updated without the need for physical intervention; previously, these features, in order to be updated/upgraded, needed a physical replacement with the consequent criticality of the disposal of the old model. The third barrier of servitization then, or the raising of entry barriers, is a variable also present in the digital transformation, as both phenomena, raising the level of complexity of the market, create an ecosystem in which for a potential competitor, which is not having the same resources or having only a purely product-focused offer, it is really hard to fit.

Finally, the last enabling effect of servitization is strictly connected with the last two of the other group: the latter, in fact, i.e. the desire to provide advanced services with great complexity, and the decision to seek greater personalization of their product, are nothing more than a direct consequence of the desire to extend the offer and make it more elastic and dynamic.

What emerges from the comparison carried out is that often the decision to adopt a business model based on servitization coincides with the development of digital capabilities, or if this is not the case, it implies its development over the long term: the enabling effects of digitalization are, in fact, coinciding with those of the PSS or are in any case a direct consequence.

The interconnection between servitization and digitalization can be analysed also by looking at contextual factors that characterize both phenomena. Indeed, we can state that both servitization and digitalization have an impact on value creation, design of the offering, marketing and delivery system, and company's value chain.

Looking at the value creation process, the acquisition of the data of the installed base, which we could define as the production process of the material with which the value of services is built, receives from the digital development a substantial improvement both in terms of effectiveness (data volume) and efficiency (made available more quickly). Therefore, as Lenka et al. (2017) also argues, it is clear how the technological innovations are inherent in digitalization, subdivided into intelligence, connect and analytical capabilities, respectively, which help to enable or significantly enhance two of the crucial mechanisms for the co-creation of the value, that is the perceptual and the responsive one, making the organization more aware of its clients' processes and making it more reactive, so that they can respond to their needs in a timely manner. As highlighted by the second factor of the portion of the scheme on digitalization, the advantages of the latter, for companies with a serviced business model, are not limited to only internal processes, but also extend to the area of creation of the value shared with the client, ensuring the possibility of increasing the complexity of the services provided to the customer and greater cooperation with it.

Again digitalization emerges as a critical factor to make servitization more effective and efficient, an element not necessary for the development of a PSS, but that if present contributes to an increase in its results.

For what concerns the design of the offering, surely the most incisive evolution in this field is the fact that the offer is developed in modules (product modules, service and information, respectively), and often their development is parallel, so as to favour their integration later. The service modules then compared to before now follow a structured development, with steps similar to those of product development (Kindström and Kowalkowski 2009). Usually, the process sees the back-office units developing product modules (Kindström and Kowalkowski 2014) and standardized service (Ulaga and Reinartz 2011) using the information gathered from the installed base and front-office units, but the latter can then be developed/ modified also locally to meet specific customer needs. This elasticity of the offer is clearly made much simpler if there are digital elements inside the products, as the offer can be easily modified even after sales with a simple software update remotely. Moreover, thanks to the development of an advanced technological platform, it is possible to obtain a considerable advantage in terms of ease of data sharing and communication between the various back and front-office units, standardizing their practices and creating a single channel through which they pass, which, especially for very structured companies, is not so obvious given the geographical dispersion of the various facilities.

The impact on marketing and delivery system can be analysed looking separately at the sales phase and after-sales phase.

Going to assess the bonds clearly we need, for both phenomena, to search for the sales phase new profiles, with diversified knowledge ranging from technical to the most economic/managerial, to not only know how to describe the offer from a point purely technological, but also be able to ride the "flow of data", interpret them and turn them into information to communicate to the customer. This factor was also present in the barriers of both issues, given the difficulty in finding such profiles on the labour market, but once obtained, they assume a primary role in the collaborative process of creating value that is established between the organization and its customers, as they are often placed in areas in direct contact with them to take full advantage of their characteristics.

As for the post-sales phase, for the servitization, the changes it implies are:

- A field service network is required (Kindström and Kowalkowski 2014) and

- Continuous relationship with the customer, sometimes even for the entire life of the product.

While for the digitalization we will have the following:

- Development of new business functions to manage this phase (Porter and Heppelmann 2015), such as those given below:
  - Dev-Ops: Function that monitors the performance of the installed base, and develops and manages the updates of the offer, both software and services.
  - Customer success MGMT: Function that manages the customer experience, guaranteeing the customer to maximize the value obtained by the company PSS.
- Ability to provide services remotely and in a preventive manner (Porter and Heppelmann 2015; Baines and Lightfoot 2013).

Finally, the last subcategory of the transformation elements concerns the changes that the two phenomena imply for the network of actors who are involved in the organization in order to produce the final offer that is offered to the customer.

As we can see, the expansion of the supply chain is a common element to both phenomena, even though as in other similar cases in the model, the meanings are slightly different in the two cases: if, in fact, talking about servitization, this event is motivated by the search to reduce the complexity and the risks of managing a PSS offer, or to increase the value of the overall offer, and in the case of digitalization, we also have a need here to find allies to deal with complexity, but, in this case, it is not about supply but rather the construction and management of a technological platform, which emerged from the analysis of the other parts of the scheme which coincides with the ultimate result of this phenomenon. Hence, there is the need to open up to new actors, with skills in electronic or IT, to support the company in this difficult task. Moreover, we are moving towards a more and more specialized specialization of companies, given the increased complexity of the activities, and this requires a great communication among all the actors: a technological platform; in this case, can greatly facilitate the practice, creating a homogeneous environment and a standard channel in which information exchange can take place.

Seven Key Facts

- The concern towards the economic, environmental and social dimensions of sustainability is shaping modern competitive contexts.
- Collaborative consumption and sharing economy offer new and unexplored chances to companies.
- The concept of redistribution, posed at the interception between sharing and circular economy, is a promising new approach to reuse and for circularity of goods.
- Servitized offerings may differ significantly in B2B and B2C markets.

- Many successful implementations of PSS can be found in manufacturing industry, but given the variety of cases, it is hard to generalize success factors.
- Examples of successful PSS implementation can be found also in sustainability-driven contexts, e.g. supply of renewable energies.
- Digitalization can be seen as a parallel path to servitization, imposing a series of structural and procedural changes.

References

A. Azarenko, R. Roy, E. Shehab, A. Tiwari, Technical product-service systems: Some implications for the machine tool industry. J. Manuf. Technol. Manage. **20**(5), 700–722 (2009)

T. Baines, H.W. Lightfoot, Servitization of the manufacturing firm. Int. J. Oper. Prod. Manage. **34**(1), 2–35 (2013)

A.P.B. Barquet, M.G. de Oliveira, C.R. Amigo, V.P. Cunha, H. Rozenfeld, Employing the business model concept to support the adoption of product-service systems (PSS). Ind. Mark. Manage. **42**(5), 693–704 (2013)

R. Botsman, R. Rogers, *What's mine is yours, the rise of collaborative consumption* (HarperCollins, New York, NY, 2010)

R.B. Bouncken, A.J. Reuschl, Coworking-spaces: how a phenomenon of the sharing economy builds a novel trend for the workplace and entrepreneurship. RMS **12**(1), 317–334 (2018)

M. Cook, T.A. Bhamra, M. Lemon, The transfer and application of product service systems: from academia to UK manufacturing firms. J. Cleaner Prod. **14**, 1455–1465 (2006)

W. Coreynen, P. Matthyssens, R. De Rijck, I. Dewit, Internal levers for servitization: How product-oriented manufacturers can upscale product-service systems. Int. J. Prod. Res. **56**(6), 2184–2198 (2018)

J. Elkington, Corporate strategy in the chrysalis Economy: Corporate Environmental Strategy **9** (1), 5–12 (2002)

S. Emili, F. Ceschin, D. Harrison, Product-service system applied to distributed renewable energy: A classification system, 15 archetypal models and a strategic design tool. Energy Sustain. Dev. **32**, 71–98 (2016)

C.A. Friebe, P. von Flotow, F.A. Täube, Exploring the link between products and services in low-income markets—evidence from solar home systems. Energy Policy **52**, 760–769 (2013)

H. Gebauer, E. Fleisch, T. Friedli, Overcoming the service paradox in manufacturing companies. Eur. Manage. J. **23**(1), 14–26 (2005)

M.J. Hannon, T.J. Foxon, W.F. Gale, 'Demand pull' government policies to support product-service system activity: the case of energy service companies (ESCos) in the UK. J. Cleaner Prod. **108**, 900–915 (2015)

M. Iansiti, K.R. Lakhani, Digital ubiquity: How connections, sensors, and data are revolutionizing business (digest summary). Harvard Bus. Rev. **92**(11), 91–99 (2014)

O. Isaksson, T.C. Larsson, A. Öhrwall Rönnbäck, Development of product-service systems: Challenges and opportunities for the manufacturing firm. J. Eng. Des. **20**(4), 329–348 (2009)

M. Khorram-Niaki, F. Nonino, *The management of additive manufacturing. Enhancing business value* (Springer International Publishing AG, Basel, CH, 2018)

D. Kindström, C. Kowalkowski, Development of industrial service offerings: A process framework. J. Serv. Manage. **20**(2), 156–172 (2009)

D. Kindström, C. Kowalkowski, Service innovation in productcentric firms: A multidimensional business model perspective. J. Bus. Ind. Mark. **29**(2), 96–111 (2014)
B. Kuijken, G. Gemser, N.M. Wijnberg, Effective product-service systems: A value-based framework. Ind0 Mark. Manage. **60**(1), 33–41 (2017)
B. Laperche, F. Picard, Environmental constraints, product-service systems development and impacts on innovation management: Learning from manufacturing firms in the French context. J Cleaner Prod. **53**, 118–128 (2013)
S. Lee, Y. Geum, H. Lee, Y. Park, Dynamic and multidimensional measurement of product-service system (PSS) sustainability: a triple bottom line (TBL)-based system dynamic approach. J Cleaner Prod. **32**, 173–182 (2012)
S. Lenka, V. Parida, J. Wincent, Digitalization capabilities as enablers of value co-creation in servitizing firms. Psychol. Mark. **34**(1), 92–100 (2017)
C. Lerch, M. Gotsch, Digitalized product-service systems in manufacturing firms: a case study analysis. Res. Technol. Manage. **58**(5), 45–52 (2015)
R. Lozano, Envisioning sustainability three-dimensionally. J. Clean. Prod. **39**(16), 1838–1846 (2008) http://dx.doi.org/10.1016/j.jclepro.2008.02.008
E. Manzini, C. Vezzoli, A strategic design approach to develop sustainable product service systems: Example taken from the 'environmentally friendly innovation' Italian prize. J. Cleaner Prod. **11**, 851–857 (2003)
A. Marino, P. Bertoldi, S. Rezessy, B. Boza-Kiss, A snapshot of the european energy service market in 2010 and policy recommendations to foster a further market development. Energy Policy **39**, 6190–6198 (2011)
D. Maxwell, R. van der Vorst, Developing sustainable products and services. J. Cleaner Prod. **11**, 883–895 (2003)
H. Mei, Analysis of products supply chain service system development and application in service-oriented manufacturing mode-taking professional audio equipment system for example. Appl. Mech. Mater. **340**, 199–203 (2013)
M.P. Miles, L.S. Munilla, J. Darroch, Sustainable corporate entrepreneurship. Int. Entrepreneurship Manage. J. **5**(1), 65–76 (2009)
H. Patzelt, D.A. Shepherd, Recognizing opportunities for sustainable development. Entrepreneurship: Theor. Pract. **35**(4), 631–652 (2011)
M.E. Porter, J.E. Heppelmann, How smart, connected products are transforming competition. Harvard Bus. Rev. **92** (2014)
M.E. Porter, J.E. Heppelmann, How smart, connected products are transforming companies. Harvard Bus. Rev. **93**, 96–112 (2015)
R. Roy, K.S. Cheruvu, A competitive framework for industrial product-service systems. Int. J. Internet Manuf. Serv. **2**(1), 4–29 (2009)
S. Shokoyar, S. Mansour, B. Karimi, A model for integrating services and product EOL managemen in sustainable product service system (S-PSS). J. Intel. Manuf. **25**, 427–440 (2014)
S. Sorrell, The contribution of energy services to a low carbon economy. *Tyndall Centre Technical Report 37*. Tyndall Centre for Climate Change Research (2005)
S. Sorrell, The economics of energy service contracts. Energy Policy **35**, 507–521 (2007)
A. Tukker, Product services for a resource-efficient and circular economy—a review. J. Cleaner Prod. **97**, 76–91 (2015)
W. Ulaga, W.J. Reinartz, Hybrid offerings: How manufacturing firms combine goods and services successfully. J. Mark. **75**(6), 5–23 (2011)
V.K. Velamuri, B. Bansemir, A.K. Neyer, K.M. Möslein, Product service systems as a driver for business model innovation: Lessons learned from the manufacturing industry. Int. J. Innov. Manage. **17**(1), 1–25 (2013)
S. Witjes, R. Lozano, Towards a more circular economy: Proposing a framework linking sustainable public procurement and sustainable business models. Resour. Conserv. Recycl. **112**, 37–44 (2016)

Chapter 4
How to Trigger the Strategic Advantage of Product Service Systems

This chapter analyses the topic of servitization and PSS with a particular attention to the main strategic issues involved. It has the aim of presenting and explaining some new strategies in developing servitization, linked to recent or renewed business trends and models.

This is the case, for instance, of the sharing economy and circular economy phenomena that can be considered as a derivation of servitization.

4.1 Translating PSS into Competitive Strategy

In recent years, there have been many changes involving manufacturing world. More and more companies are deciding to focus on a market that no longer aims at the simple sale of a material product, but rather that of a real offer able to satisfy the final customer as much as possible. This change, which is increasingly moving towards a service economy, has the merit of being attentive to the environmental problems that are affecting the current world. From a past where resources were seen as inexhaustible, to a present increasingly aware of the importance of the end customer and the environment in which we live. The current problem, in order for this to take place, is that of being able to spread this concept also to people with a mentality still facing the past, where more importance was given to the possession of the material good in itself than to the objective. The service economy is seen as a functional economy, precisely because it aims to sell a certain function, where the purchase of transport rather than the vehicle is preferred, the purchase of printing in place of the photocopier.

In this context, companies can plan innovative business models to exploit all chances offered towards a creation of a competitive strategy based on servitization and on a multitude of elements that can be considered in design and development of a product service system.

A. Annarelli et al., *The Road to Servitization*,
https://doi.org/10.1007/978-3-030-12251-5_4

4.1.1 The Process of Strategy Formulation

The process of strategy formulation requires companies to define an appropriate course of action, to ensure its capability of reaching strategic objectives. It is, obviously, a key process that must be undertook in each organization in order to ensure a proper strategic planning that matches firm's goals with its resources and capabilities. This process also involves a careful study of the context(s) and environment(s) in which an organization operates. A strategy formulation process is essential to develop a valid strategy and provide the organization a clear focus and direction, translated into a plan of actions.

The process of strategy formulation can be divided into six steps that are widely acknowledged in practice and literature:

1. Study and definition of the organizational context;
2. Study and definition of the strategic mission;
3. Study and definition of strategic objectives;
4. Study and definition of competitive strategy;
5. Implementation of developed strategies;
6. Analysis and evaluation of progresses and results performed.

Defining the Organizational Context

This is the first step of strategy formulation process and the main aim consists of a clear identification and definition of the company and its activity, in relation to its customers. Indeed, a company could not succeed competitively and economically without matching the needs of its customer base.

In the specific context of strategy formulation for PSS, the importance of these needs is even more important as the foundational element of the overall value proposition on which the business model is centred. Organizations must always take into account the reasons behind customers' choice of acquiring a certain product/service offering, more precisely we can state that customers are looking for benefits rather than features, and the identification of these benefits is the first and most important part of the overall strategy definition process.

The subsequent step involves the identification of one or more target groups: this process should mostly take into account psychographic indicators more than the mere demographic ones. It is therefore much more important to identify (potential) customers according to their values, opinions, attitudes and consequently their lifestyle and habits. This is true especially in the context of B2C markets. Looking at more rational and predictable B2B markets, it is not possible to take into account all psychometric factors described above, but companies' mission and vision, for instance, can be considered a good proxy for values, attitudes and opinions.

The last key element is related to technological advance and innovation rates: these usually cause quick changes in markets and marketplaces, making competitive contexts always more turbulent and uncertain. With technological and digital innovation spreading across almost all industries nowadays, these competitive issues can be considered as a common element of mostly all contexts in which modern companies operate: in this regard, servitization can offer a significant strategic flexibility to companies, if adequately developed and inserted in the process of strategy formulation.

Defining the Strategic Mission
With the term strategic mission is usually addressed the long-term perspective over strategic goals. It is the first and key element in defining strategic paths and strategic priorities in companies' development plans. The formulation of the strategic mission requires companies to match its own values and vision for the future, with its resources, abilities, capabilities and market positioning.

Defining the Strategic Objectives
The definition of strategic objectives mainly requires organizations to identify performance targets, in accordance to which the objectives must be translated and operationalized. Examples of this operationalization effort include, for instance, production rates of goods/services, relative market position, market share, rates of customer service, rates of innovation development and introduction.

A key step in this part of the process is the communication to companies' employees and stakeholders of the set of strategic objectives and performance targets. Furthermore, the involvement of members of the organization can ensure a more proper and punctual translation of the objectives on the operational and individual level.

Defining the Competitive Strategy
Connecting the strategic objectives to the competitive context in which the company operates is the subsequent step in the strategy formulation process. It is essential to understand this competitive positioning not only for the organization as a whole but for all functions, departments and individuals. Indeed, the company's competitive position depends on the contribution of all areas and elements inside the organization.

Firms are constantly threatened by changes taking place on the marketplace: the strategic key in reacting to these changes is developing proactive responses and communicating them with the entire organization at all levels, so as to build a shared knowledge and consensus on strategic threats and responses.

Last step in defining the competitive strategy requires identifying key resources and determine how these will be used. The resource allocation requires as well the involvement of all areas, functions and department of the organization, so as to allow a full exploitation of available resources in the most effective way, to contribute in meeting strategic needs and objectives.

Three key factors must be specifically taken into account when developing the competitive strategy, and these are the industry/marketplace, the competitive position, and internal strengths and weaknesses.

The analysis and evaluation of industries must include market size, market growth rates (as emerging from historical data), potential profitability of the market, rate of new entrants and threats linked to the nature and characteristics of the industry. All these factors must be considered and evaluated in an iterative and ongoing effort of industry and market monitoring.

For what concerns the competitive position, a key requirement for companies to succeed in competition is having a full understanding of other players operating on the market. The identification of competitors demands firms to be able of understanding their strengths and weaknesses, together with the ways in which their offerings meet customers' needs.

As the last element for the definition of competitive strategy, it is vital for companies to be aware of their own strengths as well as weaknesses, and how these relate to and interact with the competitive context. An accurate management of strengths should focus on their leverage in order to maximize the deriving advantage. On the other hand, the awareness of weaknesses and related vulnerabilities is a crucial step in defining appropriate response strategies to threats and in identifying areas of improvement.

Implementing Strategy
Once the organization has defined itself, its goals, studied the marketplace and competitors, the strategy must be put in place. Tactics play a key role in this implementation step, as they can be defined as actions that enable organizations to build and develop the foundations for strategy implementation. Companies must understand the iterative nature of this process, since as long as the most (and less) effective tactics are identified, strategy implementation methods are changed and improved dynamically.

Evaluating Progress
Following implementation, plans regularly need to be monitored and results need to be measured and evaluated. As every strategic plan is organized in objectives and performance targets, it is vital for organizations to understand how the plan is put in action and how progresses are matching those goals. If requirements and goals are not being met, it is an important alert for companies that should take on changes in an adaptable and flexible way.

4.1.2 The Role of Path Dependence in the Strategy Formulation

There are no significant differences attributable to a specific industry or to firms' dimensions, as emerging from studies on PSS, but one of the main constraints and obstacle emerging is that of path dependence (i.e. firm's previous investments and its repertoire of routines, its "history", that constrain its future behaviour), which seems to highly affect PSS adoption and success.

The first step in understanding the concept of path dependence and its effects is the difference between "conventional economics" and "new positive feedback economics": according to Arthur (1990), businesses and firms that can be identified as belonging to the first category largely avoid increasing returns or path dependence, while the second ones embrace them.

The basic definition of path dependence concerns the effects of minor or apparently inconsequential advantages that can exert important and non-reversible impacts on firms' set of decisions, like for instance resource allocation. It can also be said that path dependence produces an effect of "lock-in by historical events" (Arthur 1990).

There are three different types (or forms) of path dependence, distinguished in first-, second- and third-degree path dependence (Liebowitz and Margolis 1995). The first form, namely, first-degree path dependence, arises when this sensitive dependence causes no actual harm or undesirable outcome: this is the case of initial actions that constrain a firm on a certain path (which exhibits switching or changing costs), but that appears to be an optimal path. In this case, even if some errors might have occurred in the process of strategy formulation, organizations might derive useful lesson learned and tacit knowledge that could be useful in a self-reinforcement process of strategy development and implementation.

Second-degree path dependence is strongly linked to the presence of imperfect information. Indeed, companies sometimes make efficient decisions on the bases of available knowledge, which turn to be inefficient in retrospect. Conversely, the chosen path (and consequent lock-in) in this case brings the company towards inefficient results, but the optimal and efficient path was unknowable at the time decisions were made. This situation cannot be considered as really inefficient, given the limited knowledge and the deriving boundaries. "Where information is imperfect it is inevitable that some durable commitments are shown to be inferior as information is revealed. This problem is present with any action, but is highlighted under conditions of sensitive dependence on initial conditions" (Liebowitz and Margolis 1995; p. 6). Second-degree path dependence can be seen as a consequence deriving from errors in foresight.

The third-degree path dependence, similarly with the second one, appears when the dependence from initial conditions puts the company on an inefficient path exhibiting lock-in effects, but in this case there was enough information to recognize the inefficiency. That is, there were some specific and feasible adjustments to recognize and achieve an optimal solution, but that solution was not obtained. Third-degree path dependence occurs when no arrangement is made to take into account all costs and benefits deriving from strategic and operational choices, highlighting criticalities in the strategic formulation of a company.

Case Study

Third-Degree Path Dependence in Servitization

This case presents a Product-Oriented PSS offered by a company that provides white goods in a B2B market, specifically focusing on business laundry services and food services. These services are the core of the product-oriented offering and include education, management of usage information (to reduce downtimes), financing options, maintenance and spare parts provision. This PSS does not constitute a competitive advantage and does not allow the company to outperform its competitors: services provided to customers are considered as a necessary feature of the product that cannot be seen any more as a "plus" in the offering; not offering these services would imply a failure in meeting clients' basic needs. The most important elements of the PSS considered have been recognized to be resources, mainly consisting in service infrastructure and dedicated human resources, together with organizational processes, ensuring a widespread presence on territory and an excellent quality of maintenance services. Capabilities are not considered as a key element of this PSS offering and they easy to be retrieved; the service infrastructure, despite is an important element of PSS, has been judged as easy to be replicated, while the distinctive organizational processes are part of company's tacit knowledge.

In the case presented, developing a PSS-related offering as a simple need in responding to market demands and pressures can be viewed as an example of third-degree path dependence. Indeed, developing it and investing in service infrastructures without the aim of differentiation (no will to gain a competitive advantage) put the company in an unfavourable competitive position, since this current offering did not provide any competitive advantage, and further developments of the PSS-related structure would demand non-negligible costs to the company. The choice of investing in PSS as a mere response to an immediate need, without a long-term-wise planning, put the company towards an inefficient path that could have been easily forecasted and avoided.

(*Information and data presented are took from interviews conducted by the authors*).

4.2 Traditional Strategies Driven by PSS

The concept of strategy can be seen as a right combination of objectives (corporate mission) and functional policies (guidelines, tactics), which together define the position of the organization on the market. The strategy therefore includes the attainment and maintenance over time of the objectives, rather than the pure and simple achievement of a single success; it is a continuous process that joins the evolution of the market by combining realistic objectives with assets that allow the company to achieve them. For a company to succeed, its strategy must maintain a fair balance between its strengths and weaknesses with the opportunities and threats of the external market.

The generic strategies that a company can adopt can be identified by two dimensions (Porter 1985):

- Target market (broad target-narrow target) and
- Competitive advantage (cost-differentiation).

These dimensions make it possible to identify different types of strategy: cost leadership, differentiation and focus (Fig. 4.1).

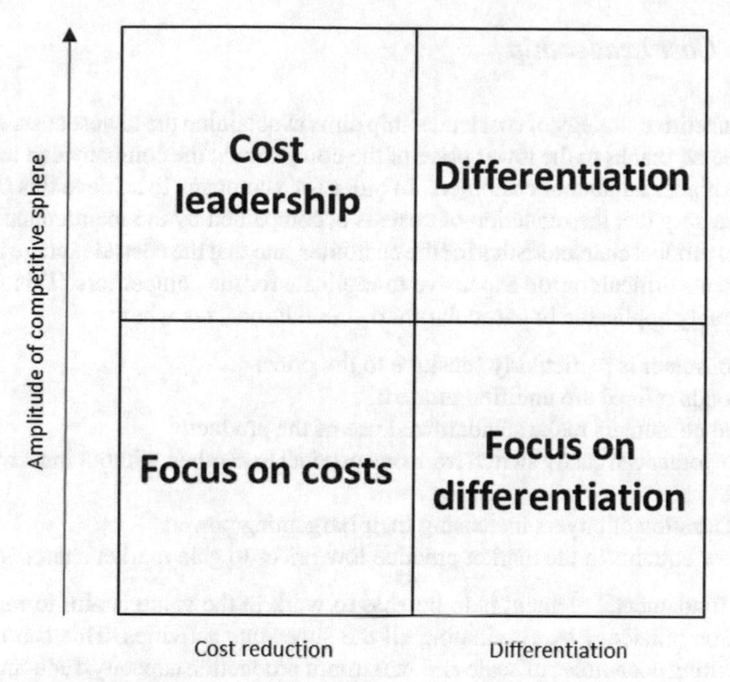

Fig. 4.1 Categories of competitive advantage

Cost leadership and differentiation represent: "[…] a fundamentally different approach to creating and gaining a competitive advantage […] normally a company must make a choice among them or remain locked in the middle" (Porter 1985). Differentiation is an expensive process and cannot therefore be achieved at reduced costs (and therefore prices). Companies that decide to adopt the first strategy aim to increase their market share by creating a product that differs from competitors for particularly low prices. To achieve their goal, there are many methods that can be adopted: large-scale production, continuous process improvement, minimization of costs and waste, TQM, benchmarking and continuous control of the external environment. On the other hand, the companies that decide instead to focus on the second competitive strategy decide to invest to make their products and services unique in the market and, thanks to this feature, we can afford to maintain a higher price than that most of the competition.

There are many studies that focus on the characteristics of the resources and processes on which a company must aim to obtain a competitive advantage and, above all, a high level of performance that is constant over time, independently of the strategy adopted. The idea is that only a company that has the ability to maintain its high level of performance over time will be able to have constant financial returns.

4.2.1 Cost Leadership

The competitive strategy of cost leadership aims at obtaining the lowest costs within a given sector, thanks to the lower price of the competitors; the company can therefore attract a higher number of customers. In order for a company to achieve this strategy, it is necessary that the reduction of costs is accompanied by the maintenance of the essential product characteristics for the customer, and that the cost advantage is based on elements difficult or too expensive to replicate for the competitors. This strategy is preferably applicable in particular market conditions, i.e. when:

- the consumer is particularly sensitive to the price;
- the goods offered are undifferentiated;
- all end consumers make standardized use of the product;
- the customer can easily switch from one product to another, without incurring high costs;
- concentration of buyers increasing their bargaining power;
- the new entrants in the market practice low prices to gain market share.

The fundamental element is to be able to work in the value chain, to maximize production efficiency by eliminating all the super-flue activities. This can be done by exploiting economies of scale and maximum production capacity, reducing costs, outsourcing or integrating activities and offering an increasingly standardized product.

With regard to the cost leadership strategy, the achievement of the objectives is based on the achievement of an operating result of efficiency; therefore, when such sources of efficiency should be temporary, easily imitated or rendered no longer functional due to the advent of new and improved technologies, the competitive advantage would be temporary and would not lead to continuous and long-term profitability. Barney in 2002 to explain this concept states: "[…] if cost leadership strategies can be implemented by many companies in an industry, or if no business is facing a disadvantage in terms of costs in imitating a cost leadership strategy, therefore being a cost leader does not generate any competitive advantage for a company". The continuous improvement in this area is fundamental but at the same time, the diffusion of technology and of always better processes make it easily imitated by the competition. Other sources to gain an advantage are the economies of scale and organizational learning developed within the company; however, many studies have shown that the barriers to entry are particularly low. In the literature, it is possible to find many authors who underline how companies that adopt this type of strategy can be blocked in a circle due to the continuous improvement of technology and a consequent reduction in costs in production processes. As an example it is possible to think of the case of Ford Motor Company which has univocally focused on the production of the Model T to reach the lowest possible cost but, at the same time, this has made the organization vulnerable to the proposed innovation strategy and started by General Motors. Precisely, this kind of problem has indeed led Ford to change his strategy over time. The initial strategy was, in fact, based only on cost leadership, and it made it possible to make mobile cars accessible to most Americans and not just the higher end of customers. The competitive advantage was based on a massive reduction in costs due to process simplification. The main problem was the one mentioned above, namely, the difficulty in maintaining this advantage over time from the imitation of competition; in 1927, in fact, General Motors succeeded in surpassing Ford becoming the largest American car manufacturer with increasingly efficient processes but, at the same time, also offering a wider range of differentiated products. For this reason, today Ford, in order to compete in the world market, while maintaining its basic cost–loss strategy, is also moving into a generic differentiation strategy.

As highlighted above, there are many benefits from the adoption of a PSS model within the company; at the same time, there are also many barriers. It is precisely these barriers that significantly curb the realities that are working with a cost leadership strategy and therefore, in line with their company policy, cannot make huge investments for a radical corporate change. A company that decides to undertake this change certainly needs to invest time and money in the re-education of staff: something very difficult to do when working with a strategy where achieving full efficiency, with the reduction of times and costs, it is a fundamental key. An automotive manufacturer, for example, always focused on proposing to the automotive market at particularly low costs for a large number of consumers, will have difficulty shifting its focus from the production of the bodywork and engines to maintenance and car sharing, things up to now outside the company. On the other hand, it will also have to invest in a specific type of marketing to do a real psychological work on customers, reinforcing the meaning of use at the expense of properties. To circumvent this type of obstacle, companies may decide to implement branding strategies that, obviously, would result in higher costs and would therefore limit the adoption of PSS as a low-cost strategy. Another point that could discourage this type of company is the average long time in which to find the first real benefits, therefore not immediate benefits and above all difficult to account for. Even greater difficulties for manufacturing companies, which, passing from the pure and sole sale of products to the sale of a product–service, also need to change the way in which costs are accounted for in the company. In services, the customer will pay only what he thinks is the true value of the service; although it is possible to have some quantitative measures (for example, the number of service units delivered or the customer's establishment), the quality of the service, which can then determine the price, can only be measured using subjective criteria, as the satisfaction of the final customer. Furthermore, maintaining relatively low economic costs by offering a product–service in the same way and at the same time trying to reduce the environmental impact of the asset is a real challenge for companies.

A company before adopting a business model such as the PSS must consider the fact that this will involve investments in the medium-long term and uncertainty regarding the company cash flows (Mont 2004).

Precisely, for this series of motivations, companies operating with a cost leadership strategy could think of as a first step to introduce initially a product-oriented PSS, which would certainly involve not so radical changes within the company and the possibility to continue to work with a standardization policy to respond to low costs for a large customer segment. The literature also highlights the substantial difference in the adoption of services by companies that produce "simple" products from those that produce complex products' systems. When we talk about simple products, we refer to those standard products aimed at the mass of customer–typical consumers and reduced attention on the part of the studies, underlining how it is probably the least indicated strategy for PSS adoption. Nevertheless, in recent years, it has been analysed how services have represented between 10 and 13% of sales also in the area of simple and standardized products. Moreover, as already underlined in the traditional analysis of the cost leadership, more and more companies are in a situation in which they go to support the differentiation of the product to their low-cost strategy: even in the context of PSS, it is possible to find this situation. Indeed, in this context, it is the addition of the service, to the standard product, which can represent a significant opportunity to create a more competitive strategy against an increasingly hungry market, where sometimes, for many companies, it is impossible to survive by limiting reduce costs and therefore the prices of the final product.

4.2.2 Product and Service Differentiation

The strategy of differentiation aims to propose to the market a product or service with characteristics that make it unique and inimitable, thus preventing competitors from proposing the same final good. Sometimes, the differentiation consists in the offer of a product/service not yet available in the market; on the other hand, it can be seen simply as a different perception on the part of the client through proper marketing activity. This type of choice is made in particular market conditions:

- presence of heterogeneous consumers who do not adapt to a standardized good;
- sensitive consumers to the brand and uniqueness;
- extensive possibilities for differentiation by companies regarding a single good/service;
- rapid technological development that offers the possibility to improve and differentiate the offer.

The final price of the product/service will therefore be above average as it will cover the additional costs used by the company to offer a differentiated product and must reflect the value perceived by the end customer.

It is therefore inevitable that the advantages achieved with a differentiation strategy are more likely to continue over time, given the difficulty of competitors in being able to imitate products and services recognized in the market for their uniqueness. In fact, differentiation is not a simple and immediate thing, but it arises, thanks to the development of specific processes and strong investments in research and development, which allows the company to develop significant resources to maintain a high level of performance over time. Companies that focus on differentiation are companies that focus their attention on the client and on the needs that this can have and, precisely for this reason, we also create a stronger link with the end customer, a factor that in turn creates the "reputation of the company". This kind of company therefore makes investments in research and development projects, in technological alliances and in the registration of brands and patents to protect itself from competition. A series of factors that determine the emergence of products/services that are more difficult to imitate and consequently more lasting relationships between company and consumers, thanks to the recognition of the brand and the company image. At the same time, however, it must be emphasized that many studies have shown that the gains achieved, thanks to this kind of strategy, are also more volatile. Over time, companies have found themselves working in an increasingly variable and uncertain market, where the costs incurred in innovation and product differentiation are not always repaid by the needs of final consumers.

Speaking of differentiation, it is therefore good to distinguish in product differentiation and service differentiation based on the good offered to the customer. The characteristics to differentiate the product are mainly the final price to the customer, the form, the characteristics, the personalization and the best quality in terms of performance, conformity, duration and reliability. On the other hand, companies that offer services to differentiate rely on ease of order, with marks, installation and proximity to the customer in the use phase of maintenance and repair.

It is therefore the competitive motivations that increasingly suggest to the producers to visualize their product as a primary resource and, the parallel development of a service, as a source of differentiation. It is, in fact, important to underline, as also the producers operating in high-tech sectors are finding difficulties to differentiate themselves in the market only, thanks to their material good. This problem certainly arises from an increasingly aggressive market, with games to re-low the price of products. The adoption of a PSS model is therefore seen in this context as the best approach to creating a competitive advantage in the strategy of differentiating companies.

It is therefore the realities that find themselves working with a strategy of differentiation and, in the specific case, of product differentiation, which are among the most involved in the introduction of servitization, just for the final purpose of being able to further advantage increase the uniqueness of their good in the market. So many companies today are traditionally based on the production and sale of products, which now use services as a strategy for differentiation in the market and, through these services, they are able to obtain most of their income. In the field of differentiation, it is also possible to underline how the service that goes to support the product increases the barriers present with market competitors. In fact, the products–services are more difficult to imitate and at the same time the contracting power of the customers is weakened, which will have more difficulty in competing with each other due to the intangibility and flexibility of the services.

A lot of research has been done in this field to identify what are the most adopted service strategies from companies up to now focused solely on the sale of products and second to understand which are the options that would allow greater growth. In particular, in 2010, Raddats and Easingwood focused on the analysis of 25 important companies from different sectors who decided to use the service as the cornerstone of their differentiation in the market. The study includes companies that sell basic products such as metals or chemicals, to companies in the field of telecommunications or companies operating in the medical equipment, transport and aerospace industry, and therefore already focused on the sale of services. In this context, various decisions can be made on how to offer a new service or support it with the offered product: first, the case in which the product maintains its main role to which services are added; therefore, companies that focus more on service and on the relationship established with the customer, while maintaining the role of the product in a central position; finally, third, companies that decide to develop real services that are less dependent on the final product.

Services that are closely linked to the products of a company that makes the sale of the product its core business are often used to create a real differentiation of the product. The added service constitutes a part of what Porter defines as the "Value Chain" with a sustainable competitive advantage that can be realized through every activity in the chain. Porter himself, when he spoke of "market differentiation", referred to companies coming from the same sector but they managed to propose a different value chain in order to compete with each other on different bases. In line with this thought, Geuber (2008) defines companies that follow this logic as "after-sales providers" that cover a wide range of different situations, from those companies that offer a simple after-sales service (adopted more by companies that work with a strategy of cost leadership), to real complete and well-defined packages of services (Product and Service differentiation).

Then, there are companies that focus more on the relationship that is established with the customer considering the "collaborative ability" as a key to the final success. This means being able to offer services that are aligned with their products but, at the same time, also know-how to look outside and understand which types of services are more aligned with the customer's operating environments. The idea is therefore that the producer tries to understand how value is created in the eyes of the customer. The products of a company can be seen as a "gateway" for a wide range of possible services throughout the entire life cycle of the product, from the determination of the initial requirements of the customer to its complete disposal. In this shift towards this new market concept, companies that work with a product differentiation strategy will therefore find their focus increasingly on the technological leadership of "customer centricity". This change is gradually becoming more radical when a market mechanism is born in which the customer pays for the final functionality of the product and not so much for the material possession of the asset (Davies 2004), the company is therefore to have different gains based on the good or bad functioning of the service offered by its product. In this case, even companies that have their foundations in purely manufacturing, with the advent of the PSS, can not only decide to support a service to the product offered (as can be a simple maintenance), but they go to offer the availability of a good. With this new concept, service is not something explicit but is implicit throughout the product life cycle; if the good does not work, it breaks, it has some problems, it will be replaced in such a way that the customer can take full advantage of the reason why the product was purchased. Obviously, this is a type of strategy that cannot be adopted when a company is working, focusing everything on efficiency and cost-cutting, as this mechanism involves taking risks that may come to the surface, throughout the life process and use of the product. There are therefore companies that are less and less dependent on products but point to their differentiation on the service offered and, in order to develop a service strategy, at the same time it is necessary to create good relations throughout the supply chain of the service; the relationships that the company is able to create are therefore potential sources of efficiency and effectiveness (Raddats and Easingwood 2010).

It is precisely those companies that have a wide range of services offered, those that can benefit most in terms of final profits, despite the risks and initial investments required. It is for this reason that companies that operate with a strategy of differentiation, both product and service, are increasingly turning to use-oriented and result-oriented service types. The aim is to reduce production costs and materials to try to focus more on quality, on cost reduction and therefore on final profits in the end. Differentiate yourself from competitors and listen more and more to the customer's voice in order to have real economic advantages.

4.2.3 Niche Strategy

The strategy of focusing is no longer generally applied to the entire market but to a specific niche of consumers. Within this strategy, it can be divided into two further currents. The former is cost-oriented, aimed at serving a restricted circle of consumers by offering a product at the lowest price compared to other competitors in the market; the second aims instead of differentiation and therefore to offer a product at a higher price but, at the same time, customized for a specific consumer standard. The strength to be able to undertake such a strategy is that the needs of this niche of end users are not satisfied by the products/services that can be found in the mass market, regardless of whether they belong to companies that aim to products of cost leadership or differentiation. As in the previous cases, it is therefore possible to think of a context that makes it possible to apply this strategy when:

- the niche is of such a size as to generate profits;
- competition for that specific market niche is not particularly high;
- firms create a relationship of loyalty with the end customer.

The main risks involved in this situation are that the market leader can develop alternative products in his offer, covering also the needs of the niche and therefore subtracting market shares.

The niche strategy is more distinct from the other two because it is aimed at a narrow market segment, with a limited number of customers. Again according to Porter, the focus can be distinguished in cost focusing (which exploits cost differences limited to one market segment) or differentiation (which focuses on particular needs of a customer-attentive to the final value of the product or service). Often this strategy is adopted by companies that are in the maturity stage and do not want to get stuck in a sealed market with no possibility of growth. Kotler (2000) defined the niche as: "A smaller group that seeks a distinct mix of benefits". Michaelson (1988), on the other hand, gave the following definition: "to find small groups of customers that can be served inside a segment" and "to offer the customer a clearly differentiated product that fills (or creates) a need". Sometimes, this strategy is confused with the term market segmentation even if it is repeatedly stated in the literature is not a fully correct comparison. A segmentation process is a top-down process where the market is divided into smaller, more manageable sub-markets. On the other hand, the niche is a bottom-up process that, starting from the needs of a few customers (be they product or of price), creates a new space in the market. This market can be described through five key characteristics: consumers with well-defined needs, consumers willing to pay the company that best meets their needs, a market that does not attract competitors, the ability to gain particular market segments, thanks to specialization and potential for growth and good profit. The most important factors of the "niche strategy" are represented by the relationship that the company builds with its customers and its reputation. This is because long-term relationships make it possible to build high barriers to entry against competitors, and at the same time increase long-term profitability, thanks to customer loyalty. Niche markets are not easily identifiable in their infancy and therefore need to grow step by step, bringing forward possible changes, market opportunities and end-user needs.

Therefore the companies that operate in a niche market can adopt much more sophisticated and personalized PSS models. As already seen in the traditional literature, companies operating in this area have as their main characteristic a much more limited market, but they have the opportunity to interface and interact more with their customers. This factor is fundamental in PSS where there is the broadest representation of the transition from interaction with the customer to a real and proper relationship; we move from selling the product or service to providing a complete solution through a long report. There are many studies that have highlighted the success of companies that relate to customers who have really identified their needs and therefore have been able to develop real tailor-made offers.

Thanks to their characteristics, the companies that operate in a niche market have the possibility to have a higher level of servitization, creating personalized products and services to a real co-design work with customers and suppliers to represent the specifications requested by the customer. This type of company is probably the most motivated in going to undertake a path of change necessary to offer the product–service mile to the customer, a change that is part of the corporate culture and of all its employees, with particular attention to those who interact with the customer in providing the service. The integrated offer proposed by them requires, in fact, continuous contact with the client, with the consequent need for a wider range of personnel exposed to this type of work (Fig. 4.2).

It is also important to underline how, in the context of the provision of a service, the possibility of misunderstanding may be greater than those which can be incurred in the sale of a simple material product. At the same time, customers of these companies are less likely to accept errors or misunderstandings precisely because they are willing to pay more to have a service that fits ad hoc with their needs. It is precisely in this situation that the relationship between the whole product–service chain is fundamental, and how the need to have a highly qualified and prepared staff in this area

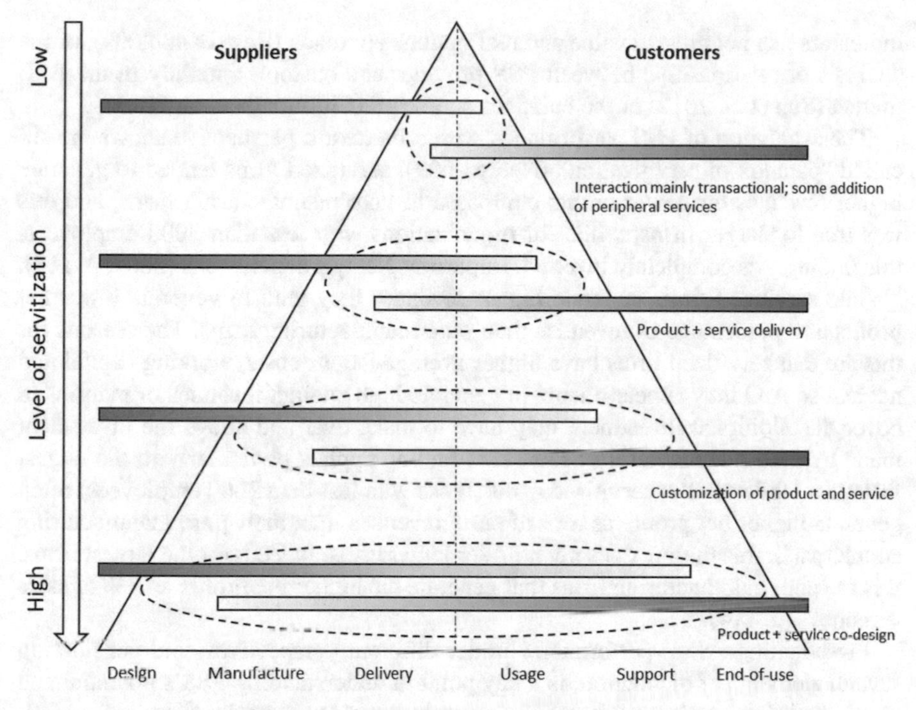

Fig. 4.2 Degrees of servitization and related strategies exemplified (Martinez et al. 2010)

is fundamental. A company that wants to provide a high-level integrated offer must constantly monitor its relationships with suppliers, with whom an intense exchange of information and know-how is required: the adoption of a new very rooted business model as it can be a result-oriented PSS is going to have repercussions on the whole value chain of the asset.

4.3 Drivers for Competitive Advantage of Product Service System

The economic potential of PSS can be related to market value for users, costs for providers, capital needs and the ability to sustain value in the future, giving the first important considerations about PSS potential linked to strategic issues (Tukker 2004).

The economic analysis of PSS can use a variety of tools like cash-flow analysis (Azarenko et al. 2009) and/or cost estimation (Nishino et al. 2012; Kreye et al. 2014), business models quantitative analysis to predict the evolution of costs and revenues' structures through years (De Coster 2011), together with a comparison of other alternatives like cost-plus and fixed-price contracts (Richter et al. 2010) or financial indicators like net present value and real options approach (Rese et al. 2009), or the analysis of relationships between PSS provider and customer, mainly in the B2B market (Sun et al. 2012) or the balanced scorecard (Chirumalla et al. 2013).

The evaluation of PSS performance with a backward perspective shows the so-called "paradox of servitization" (Neely 2009): servitized firms tended to generate higher revenues but lower profits compared to pure manufacturing firms, and this was true for larger firms; while for organizations with less than 3000 employees, this finding was completely inverted. Reporting Neely's conclusions (2009, p. 114): "While servitized firms generate higher revenues they tend to generate lower net profits as a percentage of revenues than pure manufacturing firms. The reasons for this are that servitized firms have higher average labour costs, working capital and net assets. And they appear unable to generate high enough revenues or margins to cover the additional investment they have to make over and above the investment made by pure manufacturing firms. This finding applies particularly to the largest firms, for while smaller servitized firms (those with less than 3000 employees) often generate higher net profits as a % of sales revenues than their pure manufacturing counterparts, this finding does not hold for larger firms. Indeed, for the largest firms, it is the pure manufacturing firms that generate the higher net profits as a % of sales revenues" (p. 114).

Investigating PSSs' performance under different perspectives, and considering several elements of distinction, is a key point in understanding PSS's potential and capability of generating a competitive advantage and revenues for firms.

As evidenced also by Qu et al. (2016), there is still a great need for "quantitative research to demonstrate PSS influence on society, economy, and environment", implementing "different points of view than usual ones: knowledge management, business models, technology, barriers, policy".

According to the resource-based view model, management deems some firm-specific resources and capabilities to be crucial in explaining a firm's performance (Amit and Schoemaker 1993; Teece et al. 1997). Teece et al. (1997) added another important element in the RBV, starting to consider processes as additional source of competitive advantage: "we thus advance the argument that the competitive advantage of firms lies with its managerial and organizational processes, shaped by its (specific) asset position, and the paths available to it. With managerial and organizational processes, the accent is on the way things are done in the company, or what might be referred to as its routines, or patterns of current practice and learning" (Teece et al. 1997).

The framework in Fig. 4.3 shows the distinctive elements determining the nature of PSS' competitive advantage.

At the heart of the model, there is the implementation of reuse and/or share practices and services, related, respectively, to circular economy and sharing economy: by ensuring the inimitability and protection from replication of resources, capabilities and related processes, firms should aim at defining new market segmentations and building customers' loyalty. Companies willing to servitize should focus on these drivers, while always looking for a distinctive product–service differentiation and

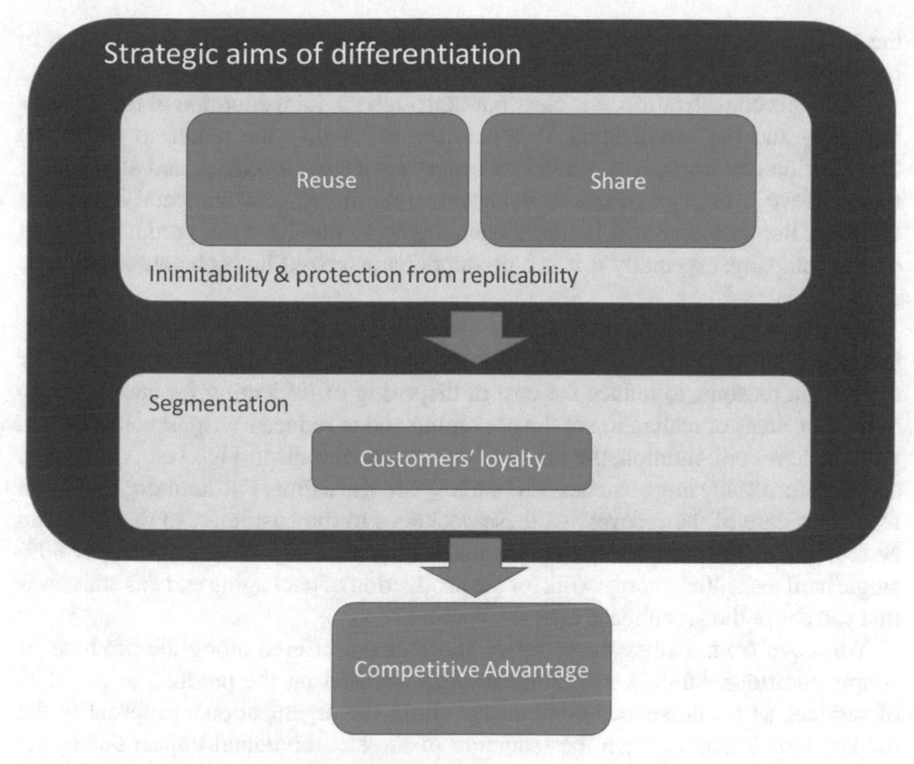

Fig. 4.3 The nature of competitive advantage for PSS

innovation, to overcome the traditional product differentiation/service differentiation distinction (Fig. 4.3).

4.3.1 *Closing the Loop: Seizing the Opportunities of Circular Economy*

To develop a product–service offering that meets the needs of the market, organizations must first have a clear understanding of the needs of the latter and the needs of the customers from which to start with development. These needs must be identified and defined systematically by the organization.

The organization will have to look at the value chain from the customer's point of view, analysing all the activities that it performs to guarantee the functionality of the product throughout its entire life cycle. In this way, services can be developed to achieve the same activities in a better way or with a lower cost for the customer or using recyclable or low environmental impact materials. This will create value for the customer, to meet also the growing environmental sensitivity of the customer in recent years.

From this consideration, it is clear how, through the joint adoption of the Circular Economy and the servitization, it is possible to identify the points in which the organization can improve to provide a better service to the client, and at the same time achieve it through processes designed to minimize environmental impact. In this way, the customer will be more encouraged to purchase the product supplied by the company, especially if it is a product characterized by high consumption or difficult disposal.

For example, the Quantum company, a hard disk supplier for personal computers, has reconfigured the packaging used for the transportation of finished products for three main reasons: to reduce the cost of disposing of packaging for customers, to reduce the costs of realization of the packaging and to reduce transport costs. In fact, with the new configuration, the filling of the means of transport has been optimized, and therefore many more batches can be transported at a time. Furthermore, Quantum also takes care of the recovery of these packages to the customer, so that they can be reused for other despatches. They found a reduction in packaging costs per 80% single hard disk, 40% energy costs for the production of packaging and gas emissions that can cause the greenhouse effect.

We move from a situation in which services are offered alongside products as simple additions within a marketing strategy focused on the product, to the offer of services as a value-added component within the organization's proposal to the market, which may concern the reduction of the environmental impact due to the implementation of business processes, the use of recyclable materials or services for the recovery of the product at the end of its useful life cycle.

In this, the circular economy supports the management of activities and processes within the network, increasing the profitability of the company, supported by a reduction in costs for the implementation of processes and/or products, and an expected increase in sales, encouraging customers to purchase the product, will see its intrinsic eco-compatibility.

For every PSS, it is important to understand if the provided services include or not practises connected to the concept of circular economy. Circular economy has been proposed as one of the latest and most important concepts to address both environmental and socio-economic issue (Witjes and Lozano 2016).

Superior performance of the PSS cases could be attributed to the presence of reuse-related services, which appears to be an order-winning criteria (Hill 1994), while basic services like maintenance, spare parts provision and educational training are considered qualifiers criteria (Hill 1994). The presence of reuse is a key element especially in the case of product-oriented offerings, ensuring superior performance in terms of competitive advantage, thanks to the environmental concern, and the series of resources and processes which make these services hard to be replicated.

Firms should involve circular economy (based on Reuse) concepts and services, so as to offer to clients a complete management of product life cycle: dismantling and take-back programmes encountered enthusiastic responses from customers, turning the durable nature of goods into an element of success, instead of representing a problem like in cases of result-oriented PSSs. Indeed, services related to the concept of reuse appear to be the key elements distinguishing cases of success (in terms of providing competitive advantage) from unsuccessful cases. Firms should invest in resources, infrastructure and organizational processes, necessary to ensure circularity in their offerings: these elements proved to be a successful and durable investment, necessary to move from offering necessary services considered as a feature and not as a plus, to providing an important source of differentiation.

Among these practices, the concept of reuse plays a significant role, since it features a whole series of product and service characteristics with the final aim of prolonging the useful life cycle of physical components. According to this, the product–service ratio presents some variations across reuse activities. With the term activities, we mostly refer to services, designed for the current users/owners of the product(s) and for potential new customers as well. In this set of activities, we can include, for instance, repair, refurbishment, reassembling, remanufacture and also redesign, with a crescent tendency from reuse towards the creation of almost entirely new products.

The framework in Fig. 4.4 shows different forms of reuse (activities) and the related degree of product–service ratio.

4.3.2 *Using Rather Than Buying: Pursuing the Sharing*

The sharing economy is proposed as a new economic model based on sharing that mainly takes place between peers, according to a peer-to-peer model. Other terms to define it are economy of sharing, collaborative economy, collaborative consumption.

Crowdfounding, car sharing, social eating and many other activities are just a few examples of this new economic model. The sharing economy was born as an alternative to the purchase of goods, replacing it with the only use based on temporary and shared access.

Collaborative consumption has become important economically only in recent years due to the emergence of new technologies, such as smartphones and geolocation systems, and the difficult economic situation. The sharing economy allows those in financial difficulty to make fruitful resources that already exist but are underutilized; it also makes it possible to use an asset that in the traditional market would not have been able to afford.

The sharing economy is a fairly recent phenomenon; it seems understandable that economic literature has not yet found a shared definition of the phenomenon.

In fact, there are many businesses that do not fully fall into the definition of collaborative economy. These are companies in which there is no longer continuous work performance but work on demand, i.e. only when there is a demand for certain goods or services. These businesses are part of a model called the gig economy.

Sharing economy is a broad and articulated concept within which a set of consumption and entrepreneurial practices converge, made possible by advances in information technology and attitudinal changes in consumers.

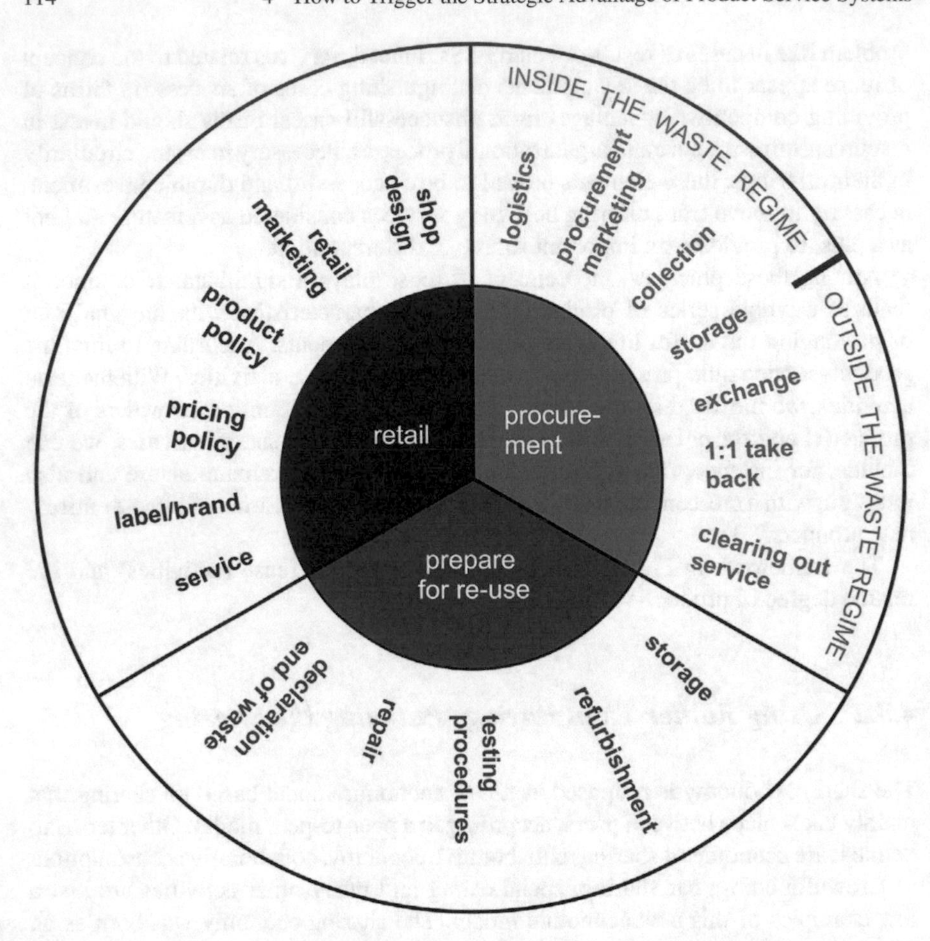

Fig. 4.4 Product–service ratio for different forms of reuse (Gelbmann and Hammerl 2015)

The sharing economy, or collaborative economy, is a relatively recent phenomenon: it was born at the beginning of the new millennium and then exploded in the years following the great economic-financial crisis of 2008. At the same time, it is a constantly evolving phenomenon, inextricably linked to the constant improvement of new technologies that allow users to be connected through increasingly refined and easily accessible online platforms.

According to Botsman and Rogers (2010), sharing economy is "an economic system based on the sharing of underutilized goods or services, free or paid, directly by individuals". In a systemic view, collaborative consumption is not seen as a niche trend or a response to negative effects of the 2008 financial crisis (Botsman and Rogers 2010), but as a new consumption paradigm joined by millions of people from all over the world.

Sharing economy is made up of so-called digital collaborative services, which connect people with other people through digital platforms (internet, mobile, tablet) that allow to share, exchange or sell products, goods and skills. These services are defined as collaborative because they foresee an exchange between peers, and digital because they are enabled by new technologies. According to the author, Internet and social media encourage the spread of collaborative behaviour for a number of reasons:

- disintermediation: the behaviour of consumers in the pre-purchase phase has changed compared to the past. People today prefer to independently collect information about a product before buying it;
- sharing: the concept of sharing in recent years has been enriched by the immaterial dimension due to the possibility of interacting through social platforms;
- confidence towards strangers: the practice of online sharing of content (of any type) by users overcomes the distrust of those who could view them and possibly use them in incorrect ways. Trust is not "blind" but in any case filtered by the social network of which these users are part and by the control exercised by it;
- living in a glocal dimension: the word glocal derives from the mixture of global and local and indicates the impacts that globalization has had on local realities and vice versa. Internet and social media allow the glocal dimension to be explored in its entirety as each individual is interconnected to others and can reach them without excessive burden.

The sharing economy is therefore proposed as a new economic model, able to respond to the challenges of the crisis and to promote more conscious forms of consumption based on reuse rather than on purchase and access rather than ownership.

Despite the many examples of collaborative consumption very different from each other, there are some points of sharing that facilitate the functioning of these realities: the concepts of critical mass, unused capacity, trust in sharing and trust in strangers.

Critical Mass

In the sharing economy, the term obviously deviates from its physical meaning: by critical mass, we mean a minimum number of assets (resources and users) necessary for the system to become self-sustainable. It is necessary as it allows the consumer to have a wide opportunity to choose and therefore more likely to satisfy their needs. For example, a bike sharing platform will have to offer a certain number of bicycles and number of stations so that the consumer perceives the alternative to traditional means of transport as adequate. The critical mass in this case has the function of making access optimal, of encouraging potential users to choose the bike instead of other means of transport, to bring them to the decision not to buy their own bicycle and that they perceive convenience and ease of use generated by the sharing platform.

Critical mass varies greatly from market to market and is very difficult to predict: it depends on particular factors such as the individual context, the needs to be satisfied and the expectations of the users. Within it, there are also a series of fixed users who use the service frequently; they will also motivate the wariest of relying on the platform, altering their habits and making a change in favour of progress. This marketing concept is called "delayed majority": after a first period in which the new product only captures the interest of innovators, over time and if innovation is positive, even the initially sceptical majority will adopt the same innovation.

Unused Capacity
One of the fundamental principles of collaborative economics is that there is an unused capacity to be exploited in some objects, characterized by a high unitary value and a low use by the user, which despite this we are used to buy. These products, not used to their full potential, in any case, involve the costs for the purchase, maintenance, repairs and finally the costs that must be incurred to buy the new version of the product. Sharing would therefore be a way to exploit and make useful the whole useful life of the asset by redistributing it among more people. This concept is not only valid for tangible goods but also for intangible assets such as time, intellectual abilities, space or goods such as electricity.

Trust in Sharing
Another fundamental principle is the belief and trust in the existence of common property that can be managed and used in sharing among all the members of the same community. Community ownership is possible through the balance between the personal interests of each individual and the interests of the community.

Trust in Strangers
Platforms, often characterized by peer-to-peer transactions, require an initial level of trust towards outsiders. Although it may seem difficult to monitor and control all the people involved in the system so that transactions are successful, the mechanism is self-governing. The role of companies is to create platforms that facilitate self-sufficient exchanges. Confidence in peer-to-peer situations seems to be relatively easy to manage, and in many cases the sense of trust in others is strengthened.

The collaborative economy can be understood as an economic opportunity, a more sustainable consumption behaviour and a more "democratic" form of economy.

The theme of sustainability can be pursued by both the consumer and the company. The individual consumer can adopt a lifestyle and a consumption behaviour that generate less environmental impacts, increasing the awareness of their choices in the "ecological" environment. At the enterprise level, strategies and activities can be implemented that aim to develop corporate responsibility and protect the environment.

In the sharing economy model, the 5R model is fully applied: reduce, reuse, recycle, repair and redistribute. Access and sharing rather than possession have the potential to more efficiently exploit resources and reduce waste. The prolonged life of resource-intensive products and the extension of use optimizes the consumption of resources and redistributes those already used.

The desire to engage in sustainable consumer behaviour plays an important role in the decision to participate in sharing platforms and activities.

Sharing is an activity that reduces environmental impacts, improves social and economic dynamics, optimizes and preserves resources for present and future generations.

The consumer shows great interest in "green" products and services and consequently is led to invest time and energy in search of a more sustainable alternative.

Although the factor of sustainability is important, often greater importance is given to the economic benefits compared to the environmental ones, and this means that the use of products and services sharing takes place because they represent a possibility of saving/earning. Only at a later stage are the environmental and sustainability advantages recognized:

- more efficient use of resources,
- waste reduction and
- less pollution and less environmental impact.

4.3.3 The Need for Differentiation and Innovation

As reported by Judge and Douglas (1998), "the natural environment sometimes offers significant new business opportunities. Some firms are discovering that by modifying the inputs, throughputs, and/or outputs of their systems, they can differentiate their goods and services from the competition and thereby gain a competitive advantage". Following the key principle that "strategic planning can and should have an impact beyond the financial performance of the firm" (Judge and Douglas 1998), for each PSS, it is important to investigate and understand reasons behind its implementation.

What emerges from companies is that firms employ a PSS-centred business model for two main purposes, which are differentiation, meaning that PSS has been perceived as a source of differentiation from competitors and a possible disruptive innovation, and necessity, which means that PSS was developed simply to adapt to competitors and/or because it became a necessary feature (i.e. extra services in the case of product-oriented offerings).

Case Study

Xerox and Conduent Inc.
When Differentiation Matters

The photocopier market is one of the most involved in servitization and is often used as an example to support the theory regarding the scope of the product service system. In particular, Xerox has been considered as a pioneer in the implementation of a "pay-per-use" formula in the printers and copiers sector. The first step still in the '90s was to move from the sale of photocopiers to that of copies, a passage triggered by the fact that the high-tech photocopiers proposed were becoming too expensive for the large segment of potential customers to whom they wanted to propose. Later, Xerox decided to go further and to propose its customers to outsource printing processes trying to become the most important service provider related to document management.

Xerox Corporation is one of the largest manufacturers of printers and photocopiers, born in the early 1900s in the United States and today present in 192 countries worldwide. Today, the company is known worldwide as a leader in the field for both knowledge and technology and is increasingly doing its strength. Over time, in fact, Xerox has gone from being considered a manufacturing company producing photocopiers to a company that supports the sale of products and the management of business processes. In this case, the level of service offered to customers is much more pushed than what can be found in the field of cost leadership and, precisely for this reason, the transition that has led the company from a vision focused only on the product to a focus on product–service, it has been long and demanding. The company in its growth has been driven by the desire to satisfy the customer and, precisely for this reason, it was the company that first started selling copies to customers rather than just selling photocopiers. Most likely, it was this spirit of adaptation and continuous search for the best solution for the increasingly demanding requests of customers, which allowed greater differentiation and therefore survival in the market. Precisely, this change of his to find new life, in a saturated market, has meant that since the end of the 80s the company was renamed "The Document Company", positioning itself as a supplier of solutions within the life cycle of the company document. So, in addition to photocopiers and printers, the company has begun to focus on new digital printing technologies and real document management services.

As outlined above, in the analysis of the literature, changes in the product service system do not always lead to immediate feedback, especially with regard to profits. In fact, even at the beginning of the 2000s, Xerox had to deal with billions of dollars invested in new ideas and technologies without immediate feedback, especially since their name remained linked to the simple photocopier company. To be able to compete, Xerox has therefore outsourced most of its basic production to focus on high-volume devices and more specialized operations that underlie its differentiation in the market. These changes meant that as early as 2010, 70% of Xerox revenues were created right in the after-sales phase of the product and the service business grew by 18% a year. Obviously, this change has been so radical that it could not happen from a simple accident but through a careful business planning, also through the recruitment of staff more and more inclined to see the importance of the final customer. It is precisely this customer solutions and services approach that now differentiates Xerox from its competitors rather than the technology itself. For many years, the company has been offering products that are not only sold but sometimes rented under special contracts that guarantee customer satisfaction through continuous monitoring of the machines and a price linked to the number of copies made. In recent years, they have also developed a document management service. The services offered today are many and range from printing consultancy, document translation, software and customer support services. From the annual report of Xerox already in 2013, we can read: "In 2013, 84% of our total revenue was based on annual annuities, which include contracted services, maintenance, supplies of materials. The remaining 16% of our revenue comes from the actual sale of equipment. [...] Our annual revenues benefit significantly from the growth of services". In the same year, President and CEO Ursula Burns told investors "[…] the transition to a driven portfolio of services is bearing fruit" and, for this reason, in 2016 the company decided to split into two distinct listed entities. On the one hand, the "Xerox Corporation" that deals with document technology, which supports the sale of products increasingly driven to satisfy customers, on the other "Conduent Inc" which bases its business exclusively on servants and, in particular, deals with of business process outsourcing.

On the one hand, Xerox retains its original name as an internationally known brand for revolutionizing the market, first with simple photocopies, and then with the addition of digital technology, software and services. The company maintains the sale of products for the home and office, but alongside a series of services that differentiate it from the competition. It is easy to understand how the company points to selling real integrated offers; for instance, the area related to the sale of printers is also entitled "solutions" and not with the simple term products. Today, the company offers the possibility to instal apps directly in printers to make easier the management of documents to print and the mobile printing service that allows printing from all mobile devices connected in the network through non-branded printers. Therefore, besides the "rental" of photocopiers, the service of supplying spare parts, maintenance and recycling of used parts, the company is increasingly trying to differentiate itself in the market thanks to the offer of "Document Management" services. One of the most important services currently offered is that of the managed press which concerns at the same time the prints, the consumables, the way in which the documents are used and the management of the processes that deals with all these aspects. The company aims to help companies in protecting their sensitive data, and printers are, in fact, a weak point through which companies can be attacked and robbed of their sensitive data. For this reason, a service is offered that can detect other connected devices and guarantee the protection of data and documents. The idea of Xerox is to offer these managed print services both to strengthen security, to optimize work procedures and to reduce the resources employed. In fact, we know that the concept of PSS is very close to that of sustainability and, thanks to this greater control of the prints made, it is possible to drastically reduce the number of copies printed unnecessarily in large companies, limited to those actually necessary and reducing waste to a minimum of paper and ink.

On the other side, it is possible to find Conduent Inc. whose name is inspired by the company's ability to connect customers in market sectors such as customer care, transport, health care and service delivery. This company was founded after the demerger from Xerox in 2016 with the aim of differentiating itself and becoming a leader in the business process services sector, maintaining those values of innovation, diversity and business integration that are at the base of the Xerox company's culture.

The company has more than 93,000 employees in 40 different countries in the world and today aims to differentiate itself in the market, thanks to the services offered as

- leadership in business process servants, thanks to solid customer relations and the proposal of differentiated solutions in growing sectors such as transport and health;
- the continuous search for final customer satisfaction through continuous investments in innovation and the development of new technologies that improve business processes;
- management of processes and users on a large scale, thanks to the differentiated range of offers proposed.

(*Information and data presented are took from the companies' websites Xerox.com and conduent.com*).

4.3.4 New Market Segmentation

The presence of different segments indicates the presence of different groups of customers with very different ideas regarding the product's property (Tukker and Tischner 2006) on the basis of which there are cultural differences and established habits (Manzini and Vezzoli 2003). A practice used to define the segments considers the different habits of the customer, since the PSS introduces changes in terms of ownership, responsibility, accessibility and costs. Focusing on the right segment with the right value proposition is a crucial factor for the success of the PSS (Kindström 2010), in fact, not all value propositions are adapted to all customers (Rexfelt and Hiort af Ornas 2009). In PSS, an effective value creation is achieved and there is an adaptation between customer and supplier BM (Nenonen and Storbacka 2010). It is problematic to choose the target segment to the same extent that it is difficult to understand the customer's perception of value and how its pre-existing characteristics influence the value proposition (Reim et al. 2015). For this reason, in analysing the restructuring of the segment, it is preferable to analyse two parameters: the first will concern the ability to understand the needs of customers, and the second will concern the targeting, or the selection of the segment on which to interact with its offer.

In PSS, since value creation must be understood through the customer's gaze (Davies 2004), it becomes critical to achieve an excellent understanding of the client, of his business and its operational activities (Kindström 2010; Reim et al. 2015). Consequently, the company should collect and analyse data and information, regarding the problems of the client and its operating activities in order to create and transmit a clear formula of value that meets the real preferences and needs of the client. Furthermore, once the needs of a client are understood, the company can influence them (Payne et al. 2008).

Furthermore, the company must develop a specific strategy for the segment, including business objectives. For this, different methods can be implemented to segment and analyse customer needs. In particular, in the PSS context, companies need to develop specific value proposals for each client, which are therefore unique (Storbacka 2011). For this reason, the company must define its market with its segment and its customers (Storbacka 2011). As seen previously, the criteria for segmentation are based on customer behaviour. This means that it is possible to segment the market according to the three categories of PSS already seen (Fig. 4.5):

- Product oriented: This segment includes all those customers who wish to remain in possession of the asset, even at the cost of major disbursements in financial terms, but who have decided to outsource the maintenance part. Consider, for example, a company that owns plants for the production of electricity that decide to buy the main components such as heat exchangers or turbines, but prefer to entrust the maintenance to their producers.
- Use oriented: In this segment are included all those customers who do not want to sustain a high initial outlay, even at the cost of not being the owners of the asset. Think, for example, of all the companies that lease means of transport or instruments of various kinds. Among the incentives that push a customer to place themselves in this segment, above all for the Italian context, there is the possibility of VAT benefits.
- Result oriented: In this type of segment are included all those customers who are not interested in owning the good or using it, but to exclusively guarantee the results produced by it. In this way, it is possible for the client not to have fixed assets related to the asset on the balance sheet.

4.3.5 Inimitability and Protection from Replicability

Together with the nature of elements constituting a competitive advantage, another important point is about their sustainability over time, and this aspect is closely related to the concepts of replicability and imitability, as exposed by Teece et al. (1997): the first one involves the "transferring or redeploying" of capabilities from an economic setting to another; the second one is simply replication performed by a competitor. According to literature, replicability and imitability are usually employed as an index of threat to competitive advantage's sustainability over time.

Indeed, even if every company can develop its own set of capabilities, competences and resources that may ensure a favourable competitive position (i.e. competitive advantage), they also must be difficult to imitate (Teece et al. 1997). Whether competitors might easily replicate or imitate a distinctive capability on which the competitive advantage is built, the whole set of resources, routines and skills related to this can result in losing its distinctive value.

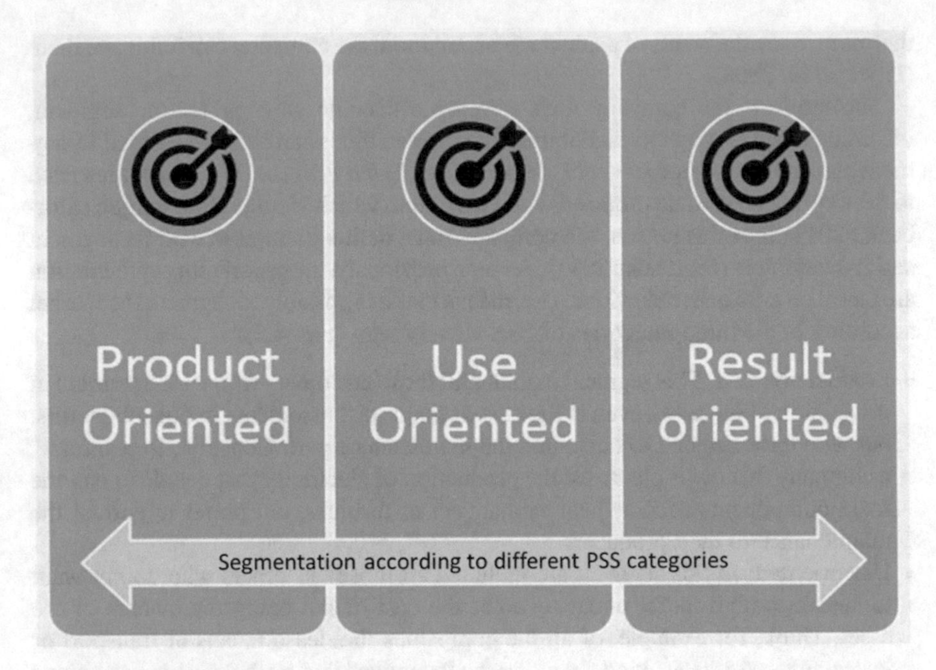

Fig. 4.5 Schematic view of PSS segmentation

However, specific sets of capabilities and competences, and routines upon which they are built, are often quite hard to be replicated: the replication effort, indeed, brings with it a double difficulty that involves, first of all, the identification of relevant routines, and then their actual replication. Furthermore, many routines, capabilities and competences are attributable to firms' specific contextual factors, like, for instance, local or regional forces that shape companies' capabilities over time (Teece et al. 1997). Another key element that can explain company-specific capabilities, opposed to local and regional forces, is the role of firm-specific history (Nelson and Winter 1982) that might affect and shape the set of a company's strategic choices, originating path dependence phenomena.

4.3.6 Loyalty as a Measure of Success

Customers' loyalty is a very important factor in determining PSS's success and affirmation as an effective market proposition. For companies, it is essential to cope with customers, since PSS has always been acknowledged as a win-win strategy for firms and stakeholders (Mont 2002).

Furthermore, acceptance from customers, represented by customers' loyalty, is a determinant of PSS success especially in the case of use-oriented models, proving that this can be considered the most attractive and interesting category of PSS, also in relationship to the considerable interest that there is nowadays towards sharing economy and collaborative consumption models.

Comparison between cases of use-oriented PSSs shows that this particular offering is experiencing a good answer from market and customers, though there are many different offerings (bike and car sharing, and co-working) from different companies with different aims. Thus, it can be stated that the use-oriented formula, based on concepts of sharing/leasing/renting and pay-per-use payment systems, is probably the case of PSS implementation that faced the most encouraging answer from the market. Instead, product-oriented and result-oriented offerings showed non-uniform results in terms of competitiveness, depending on some characteristics that are specific of each offering and business model.

Firms wanting to exploit possibilities linked to sharing economy concepts through use-oriented business models should focus on users' requests, so as to build the offering around the core needs: as evidenced, the most important element for a successful implementation of this type of offering lies in meeting customers' needs, independently from more specific characteristics like pricing, the adoption of renting and/or pay-per-use formulas and type of available products (e.g. cars/bike/scooter in case of sharing mobility). Following evidences from literature and from market, sharing mobility constitutes the most interesting and developed example of use-oriented PSS.

4.4 Evaluating Sustainability of PSS Competitive Advantage

4.4.1 Analysing PSS Risks

In the new context, it is necessary to define the risk components and analyse them as they are shared among the various actors involved in the new business model (Tukker 2004; Meier et al. 2010). Moving in a PSS context implies accepting greater responsibility for client activities and therefore accepting a considerable amount of risk (Spring and Araujo 2009). It is therefore necessary to know-how to assess and mitigate the risk (Kindström and Kowalkowski 2014), and attention must be paid to uncertainty and to the risk shared with all the actors through a serious monitoring (Meier et al. 2010). The adoption of the PSS implies for the providers an increase in the share of risk that they take on (Reim et al. 2016) to lighten the amount that weighs on the customer. The latter pays an extra cost specifically to enjoy this benefit. The risk related to PSS can usually be linked to its operational phases, which can be divided into three aspects (Table 4.1):

- Technical risk: With technical risk, we mean that set of risks related to aspects of a technical nature, such as unexpected breakdowns of machinery or their components, the use of untested technologies and the increase in operational costs and the obsolescence. Uncertainty about the future performance of the PSS is a risk factor.

- Behavioural risk: This aspect has to do with the possible risks linked to opportunistic or incorrect behaviour on the part of the client, the non-respect of the contractual clauses or the incorrect use of the solutions offered by the client that can lead to breakage or excessive wear of its most urgent components.
- Risk in delivery: Risk in managing the operation of the solutions offered, reactivity of the company in responding temporally to the needs of the customer, limits of production capacity and execution of the processes during the assembly of the solutions.

For the company to manage them in the most appropriate way, risk management is a fundamental component of business management. It consists of the continuous activity of identifying the risk, identifying the response and monitoring the results. To face the risks, therefore, four modalities are outlined (Table 4.2; Fig. 4.6):

- Risk avoidance: With this approach, we try to make sure that the risk does not occur. Choosing this option, however, reduces the supply of PSS solutions offered, given that only solutions in which you are not responsible for breakages would be offered. For example, it may be appropriate to exclude some customers or markets when the risk in those contexts would even be fatal.
- Risk reduction: The reduction of risk includes those activities aimed at reducing the severity or frequency of adverse events. This goal can be achieved by improving quality, managing information better. Risk reduction also translates into an increase in resources such as spare parts or maintenance technicians.
- Risk sharing: With this approach, you decide to spread the risk with the other actors of the network. It is one of the most interesting and used methods since it allows to spread any financial losses on multiple players and to make them less impactful. To define how to share the risk, as we will see later, the contracts assume great importance.
- Risk retention: It consists in the tendency of the providers to take on all the possible risk components with the aim of requesting a higher economic premium. This award is the one that most attracts companies that enter the PSS context (Tukker 2004).

Table 4.1 Classification of risks

Risk category	Technical risk	Behavioural risk	Delivery risk
Characteristics/ examples	Breakdowns	Opportunistic behaviour	Excessive delivery time
	Untested technologies	Incorrect use	Production capacity limits
	Obsolescence		Executive limits
	PSS performance uncertainties		

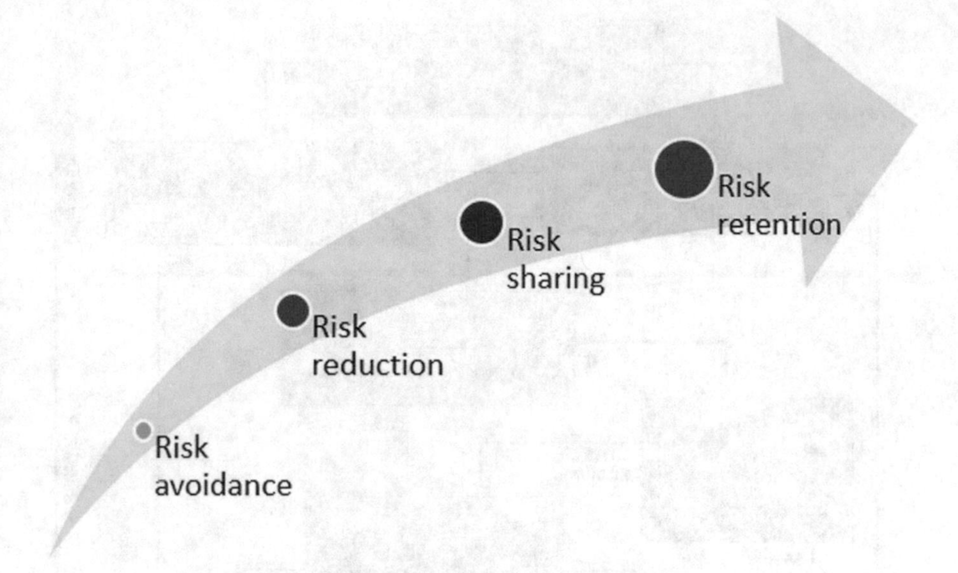

Fig. 4.6 Hierarchy of risk management strategies

All these considerations can be condensed in an evaluation and support tool for decision-making, in a prospective way (when deciding to invest or not into an alternative), but also in the day-to-day management. Figure 4.7 reports an example of a PSS risk management decision tree, where risk categories represent various levels/stages of decision-making, and a series of well-defined criteria brings to one of the risk responses analysed above.

Table 4.2 Risk management strategies

Risk management strategy	Risk avoidance	Risk reduction	Risk sharing	Risk retention
	Demobilization from contexts in which there is an excessive risk component	Quality improvement	Incentives for risk sharing	Premium risk
	Supply reduction	Proactive maintenance	Contracts of uncertainty sharing	
		Careful management of information systems		

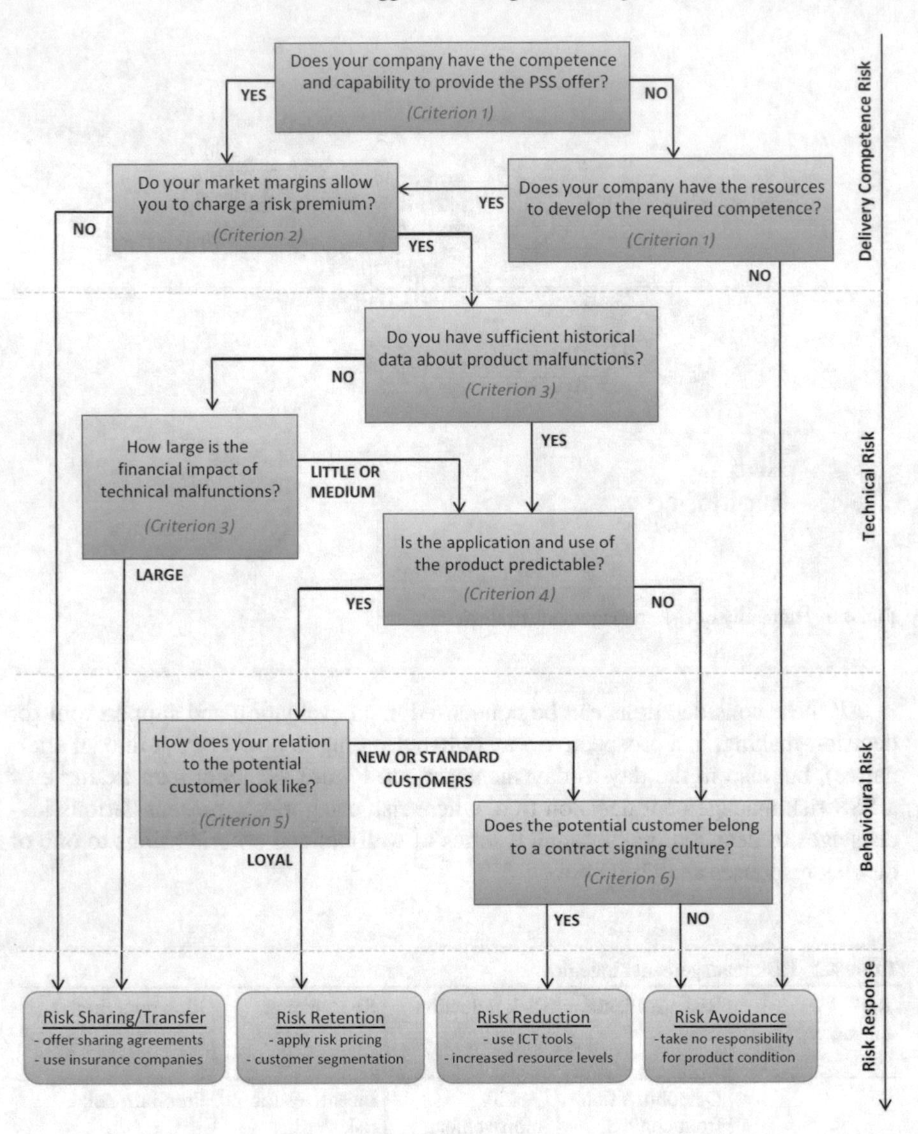

Fig. 4.7 Hierarchy of risk management strategies (Reim et al. 2018)

4.4.2 Assessing PSS Sustainability

The framework in Fig. 4.8 contains an exhaustive list of elements to be taken into account in the evaluation effort of PSS' sustainability. In evaluating the sustainability of competitive advantage, there are some key elements to be considered: first of all the price of the product, because it affects customers' willingness to pay for PSS, and also because this element is at the core of revenue flows; then another important element is cash-flow system, since PSS, as a complex business model requires a careful management and attention to cost structure. Further, at the core, there is the concept of added value that in the context of servitization and PSS assumes a whole new meaning, as already explained in Chap. 2.

Figure 4.9 presents a framework (Chou et al. 2015) where an extensive list of elements is considered in the evaluation of the overall PSS efficiency. In the model presented, a series of factors are related to customers' and employees' perceptions to evaluate the overall PSS value, while customer impact and company impact determine the overall sustainability impact: this is deconstructed in finance, resources consumption, and living conditions of customers and working conditions of employees. Both measures involved are divided into two perspectives concerning the customer- and employee-related dimensions, proving the key role that these two categories play when dealing with the process of PSS adoption. It is vital for companies to understand the key role played by these two categories, since underestimating their impact can be the most important determinant causing failures in the subsequent PSS development process.

Added Value

A company's financial performance is expressed by the Economic Value Added (EVA), which is determined by deducting the cost of capital from the operating profit, with an adjustment for taxes. As an effective and synthetic way of expressing the true economic profit of a company, it is regarded as a key economic and financial indicator to evaluate the ability of a company in generating profit and richness.

At the same time, it can be adopted also as a support in decision-making when evaluating investments, as better explained in Sect. 4.4.3.

Cost Structure

Cost structure management, price definition and consequently the revenue flow system are core activities in the PSS. The new logic that underlies the creation of value requires management to shift to pricing techniques linked to the value that is contained in the package of products and related services (Grönroos 2011). Financial and reporting activities require a review as, in terms of time, cash flows are subject to a sharp expansion, as the transfer of value no longer ceases with the delivery of the asset, but takes place over a long period of time, which obliges the provider to supply adequate financial coverage (Mont 2004). Accounting practices are therefore redefined and adapted despite applications in several contexts in the literature (Meier et al. 2010; Reim et al. 2015). Traditional assessment procedures in investment planning or cost management are no longer sufficient, since the time horizon changes (Neely 2009; Richter et al. 2010; Storbacka 2011).

At this point, it is spontaneous to ask how to build a cost structure capable of operating in the PSS context. The traditional structures, as seen previously, risk to not be enough. They are essentially three: the activity-based costing, the time-driven activity-based costing and the process-based one. Related strengths and limitations of these three methods are summarized in Table 4.3.

Given the speed with which market contexts change, the ease of adaptation of one of these systems can be decisive in the choice of their use. Companies that intend to

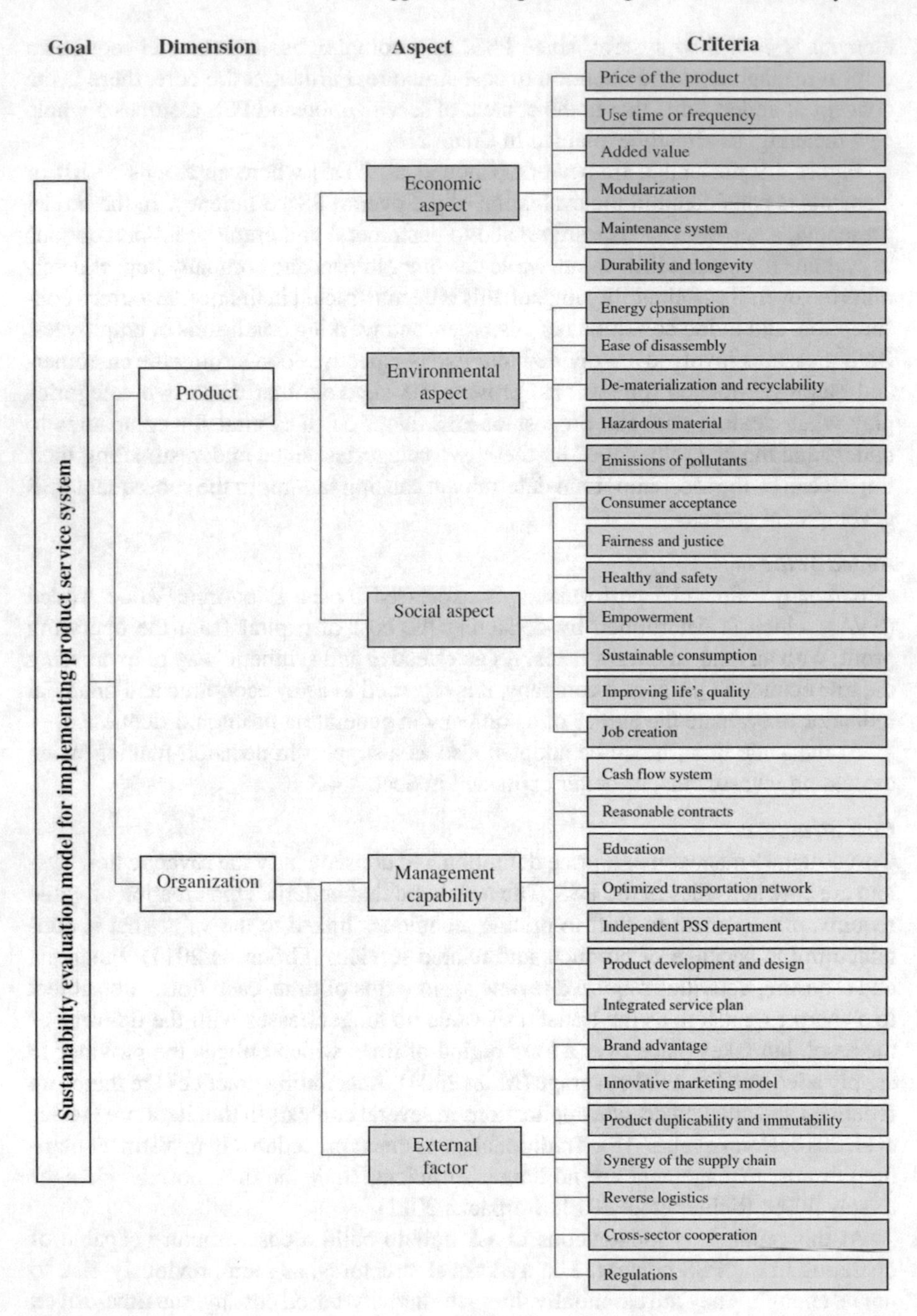

Fig. 4.8 Evaluation of product service system's sustainability (Hu et al. 2012)

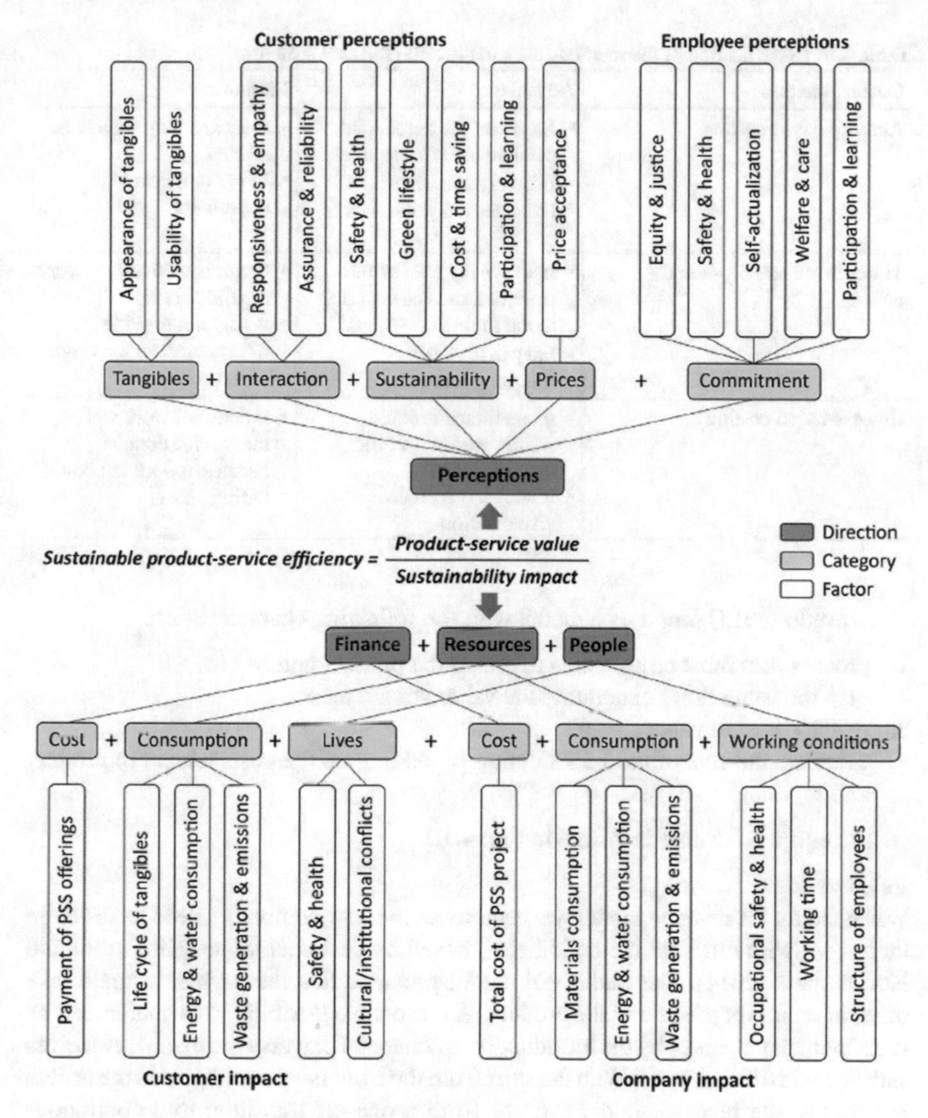

Fig. 4.9 Multiple criteria hierarchy in PSS efficiency evaluation (Chou et al. 2015)

operate in a service-oriented context must necessarily rely on a system capable of responding to key issues:

- it must be clear in expressing the costs, time and effort necessary to build, implement and maintain the solutions to be offered to the market;
- it must consider the good throughout his life cycle;
- it must be easily updated in view of the continuous updates proposed by the customer.

Table 4.3 Classification of the methods used to describe the cost structure

Costing method	Advantages	Limitations
Activity-based costing	• Based on the actual and detailed use of resources • It is very accurate • Indicates the potentials of profit	• It takes a long time to be drafted • Very complicated to implement
Time-driven activity-based costing	• It shows the cost of the activities and the cost due to the timing • Easy to update • It is very accurate	• It works well only in very repetitive tasks • It may not provide information on the origin of the cost
Process-based costing	• It can return reliable results even with little data available • It allows the processes' visualization	• It does not work well in the service context because it does not count indirect costs

Azevedo (2015) proposes a model with the following characteristics:

1. process data must be collected to create the product and service;
2. map the value flows generating the value stream map;
3. match a cost to value streams;
4. calculate the cost of the PSS solution by adding up the costs related to product and service.

The entire procedure is visible in Fig. 4.10.

Revenue Flow

When facing the revenue model, we need to define how companies need to structure their sales with different methodologies, based on customer value (Kindström and Kowalkowski 2014). The traditional model provided that the payment should take place in return for the sale of the product. A reasonable flexibility was guaranteed by instalment payments or by the introduction of financial players who made themselves satisfied with the payment. With the shift from the ownership paradigms to the access paradigms, the revenue model evolves from a one-off transition to a continuous flow based on temporal continuity or based on output (Tukker 2004; Kindström and Kowalkowski 2014). Diversity in mixed payment methods is quite common in the PSS context (Van Ostaeyen et al. 2013; Rapaccini and Visintin 2015). When we analyse this component, we can establish two subcategories: the first will be related to the management of revenue streams, or how the money transfers are structured in different usage contexts.

For a company, the transition from a sales reality based on traditional concepts to a PSS context can mean an increase in revenues through a greater offer of features to present to the market (Mont 2002). There are various types of solutions that involve the management of the property and the distribution of responsibilities on it, and different payment methods exist (Tan et al. 2010). In other words, the transition to

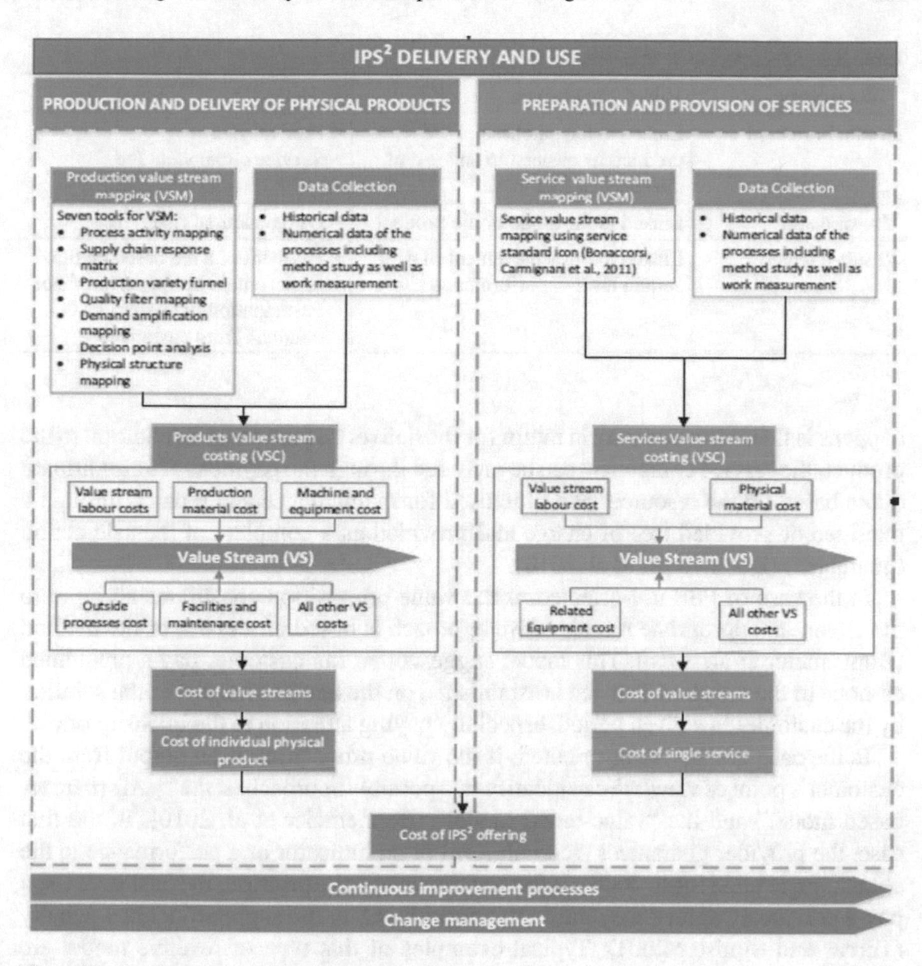

Fig. 4.10 Costing systems for PSSs (Azevedo 2015)

the PSS context makes it possible to structure the payment in different ways (Van Ostaeyen et al. 2013). Payment can be based, for example, on the availability of the product/service, on how it is actually used, on the result guaranteed by its use, thus becoming a performance-based payment method (Matthyssens and Vandenbempt 2010). So we can say that the type of revenue mechanism used is closely linked to the choice of the new proposition and depends on various variables such as the maturity of the client and the degree of projection of the company in the customer's business (Kindström 2010). Starting from the classic subdivision of the PSS seen in the production phase, it is possible to see how it is possible to set up different types of revenue models, embracing aspects of a contractual nature which will then be taken up again in the next phase and known pricing techniques.

In the typology known as product-oriented, the ownership of the asset is transferred from the asset and is transferred from the customer to the supplier. What

Table 4.4 Schematization of revenue flow

PSS category	Value proposition	Revenue flow
Product-oriented	Linked to the purchase of the product, its ownership and use of the connected services	• Product sale • Service component fee
Use-oriented	Linked to the usage of the product	• Usage method (fixed fee)
Result-oriented	Linked to the achievement of a certain level of performance	• Payment of a fee based on the achievement of the performance • Participation in re-sparks obtained from customers

happens is therefore an outlay in return for the delivery of the asset. In addition to the product, the service component can be included through the payment of an additional quota based on the resources and time used for maintenance and repairs, through a fixed fee or provided free of charge and provided as a complete of the sale action (guarantee) (Bonnemeier et al. 2010).

In the case of PSS use-oriented, if the value proposition constitutes an input to the client, an appropriate remunerative approach is based on the use of the method (Bonnemeier et al. 2010). This model suggests that the customer pay a predefined ca-none to the supplier. This fee is established on the basis of the use of the solution by the customer in a given period, especially paying attention to the intensity of use.

In the case of PSS result-oriented, if the value proposition is an output from the customer's point of view, we can identify two possible approaches: the "performance-based model" and the "value-based model" (Bonnemeier et al. 2010). In the first case, the provider guarantees a certain level of an indicator or a performance to the client (Nagle and Hogan 2006). If the supplier keeps the promise, the customer must pay a previously agreed amount. In the opposite case, the supplier incurs a penalty (Turner and Simister 2001). Typical examples of this type of revenue model are those based on the accessibility time of the machine. The second option is that of the "value-based model". The characteristic of this accreditation technique is that the customer's solution focuses on the customer's process and ensures optimization and productivity. Therefore, the price is based on the total costs that the customer saves by obtaining the solution (Sawhney 2006). In other words, the supplier benefits from the value that its solution generates for the customer. For example, in order to establish added value, total cost of ownership (TCO) analyses can also be useful in this context. It is possible to include, despite being extremely difficult to quantify, a remuneration based on the increase in the satisfaction of the downstream players of the customer who buys the solution. There are verifiable aspects related to the competition. For example, when we apply the value-based logic, the intensity of the competition forces the provider to set the parameter (for example percentage) with which to go to calculate the costs saved, or the increase in turnover (Table 4.4).

4.4.3 *Estimating the Strategic Value of Servitization*

The main aim of this paragraph is to provide a suitable methodology (described step by step) that could be replicated whenever the evaluation of a perspective servitization strategy, and markedly of its qualitative aspects, comes into play. The first step is linked to the study and the estimation of advantages and disadvantages involved in each PSS' design process. Given the high probability that these pros and cons vary from case to case, this step cannot be generalized and insightful evaluations are required from each company analysing the possibility of introducing a PSS business model.

The main steps of the methodological approach are exposed below.

- PSS scenarios are defined with reference to traditional manufacturing industries in which companies under analysis operate.
- An interview/analysis protocol has been developed and articulated on five key areas.
- Within each key area, the aspects that impact on decisions about PSS introductions are pointed out by companies' managers that have been presented with a subset of (most feasible) PSS scenarios.
- Each relevant factor is classified as advantage or disadvantage.
- According to the relevance of factors, a quantitative index is built as a reference term to discern whether, overall, advantages outnumber disadvantages, which can support decision-making.
- At the operational level, a transformation of the above index is proposed that can serve as a multiplier of the profit margin that is expectedly associated to the candidate PSS-oriented business model.

Factors Involved in PSS Implementation

In order to lay bare potential strengths and weaknesses of PSS implementations, the first necessary step consists in defining alternative PSS scenarios. Of course, these can differ (at least) according to the discussed categories (product-oriented, use-oriented and result-oriented) that commonly characterize the implementation of PSS policies. Accordingly, scenarios have to be formulated for each case. With reference to their current business, all the alternative scenarios underpinning the three categories of PSS should be elaborated. Through unstructured interviews and a large room for dialogue, companies can analyse each scenario and consider benefits and pitfalls ensuing from their potential implementation. Consistently with the research goal, considerations about potential profits or financial difficulties might be omitted.

The involvement of companies in this phase of the study allowed the authors to identify key areas of investigation, which clearly emerged from the open discussion and are listed in the following:

- Technical and design considerations: This includes, for instance, design issues, players participating in the design process, products life cycle, considerations on products' components and their (re)usability.

- Market response: Considerations linked to PSS' market potential, its capability of attracting new customers and enlarging the actual customers' base.
- Organizational aspects: This part is mainly concerned with the analysis of needs for reprocessing and/or reconfiguring existing production plants and resources, together with the possible need for new resources, competences and skills.
- Considerations on price changes and effects: This is mainly focused on the actual customers' base, and considerations here are linked to price sensibility, customers' loyalty, and how changes in price structure will affect these elements.
- External environment: Considerations on factors like suppliers' bargaining power, sensibility to price change, bargaining power of customers, competitors, incumbent firms in the market and new potential entrants.

These items represent strategic areas, or more practically, lenses under which the multi-faceted PSS phenomenon can be decomposed and analysed in its elements. All involved companies (in the pilot study) agreed on the need to investigate these areas in order to address advantages and disadvantages that might manifest when introducing PSS business models. As the spectrum of formulated scenarios ranged from promising to poorly plausible PSS implementations, it was deemed that the mostly affected business areas could be considered exhaustive. Literature mainly agreed upon main benefits and barriers coming into play in PSS adoption process (Annarelli et al. 2016): one of the aims of the study of this method was schematizing and gathering those aspects so that they could be analysed in an easier way. The individuation of the recurring key areas facilitated the articulation of an interview protocol to be used in the analysis of PSSs. In addition, instead of presenting a scenario for each of the three alternative categories of PSS, it could decide to analyse only the most promising and feasible scenario(s).

In this step, it was deemed necessary to let companies (and managers) reflect on the above areas but major specifications in terms of recurring aspects or examples. Thus, it was believed that a semi-structured form of an interview was capable of addressing the discussion towards a large number of positive/negative repercussions of a PSS scenario without any manipulation of the companies' standpoint.

Therefore, consistently with the identified key areas, the interview protocol (or analysis protocol) is divided into five sections to guide companies and managers and to make emerge their point of view and perception about PSS strategies that they have never considered before.

Of course, managers are in charge of defining whether the described PSS scenario would affect their business in the designated areas. More in details, they are expected to define which aspects, within each key area, could be affected by the proposed service-oriented scenarios.

In this context, a so-structured approach of analysis allows companies to identify a list of factors affecting PSS development and introduction, with a satisfying degree of detail and completeness. Subsequently, managers have to classify as advantages or disadvantages the factors that were found to affect the possible PSS introduction and that emerge from the analysis of the as-is situation.

In particular, factors can be classified through a matrix (Table 4.5) that assigns priorities according to effects coming into play in PSS introduction. Factors are therefore categorized according to the degrees of two variables, obtaining a set of four types. The first one represents the degree of alignment to corporate strategy of a specific advantage/disadvantage; the second variable (Table 4.5) accounts for how much a factor can impact on firms' activities and projects.

This step resulted in a key point for the quantification of the identified factors, given that each advantage and/or disadvantage can be identified as more or less relevant for the company and to what extent it can be affected.

According to this categorization and the internal/external nature of factors (compared to firm's boundaries), a scale of priorities was defined. Quantitative scores of strategic priority have been proposed for each category, as shown in Table 4.6. The assignment of the scores is arbitrary, although based on common sense and substantially agreed by the involved partner companies. To our best knowledge, no standard practice exists that assigns weight to different factors coming into play in business model shifts. Moreover, the transformation from qualitative to quantitative indicators, although mathematically and statistically not rigorous, has been employed also in some of the most acknowledged decision-making frameworks concerning PSS, e.g. Dimache and Roche (2013). Higher scores were given to external factors, because of the difficulties in coping with external environment and the assumed major magnitude of possible (positive or negative) repercussions that might follow a PSS introduction. Indeed, internal aspects were considered a minor source of obstacles. The major problem that could emerge from this category proved to be resistance to change, which is a widespread but also, supposedly, an easy-to-cope-with issue, as already remarked in PSS literature (Annarelli et al. 2016). Each time an aspect affects positively (negatively) the introduction of a servitization strategy, this is considered as an increase of advantages (disadvantages) in terms of the corresponding score. As already pointed out, just involved firms can establish aspects affecting their business. This enables a rough estimation of advantages and disadvantages as for non-monetary aspects of PSS implementations. By assigning to each advantage a "plus" and to each disadvantage a "minus" sign, it is possible to evaluate the Strategic Advantage (SA) of PSS introduction as sum of internal factors of each PSS scenario, which will be exploited as the main input for the formulation presented in next section. While SA addresses internal factors, the sum of external factors has been defined Competitive Context (CC), and their sum (giving a measure of the total impact of internal and external factors) was called Competitive Advantage (CA). This distinction was a key point in the formulation since internal advantages/disadvantages are closely linked to a whole set

Table 4.5 Factors' categories

Strategic alignment	Impact	
	Low	High
Low	Minor	Significant
High	Focused	Critical for success

Table 4.6 Factors' hierarchy and related score

Factors	Score
Critical factor for success (external)	1
Critical factor for success (internal)	0.9
Significant factor (external)	0.7
Significant factor (internal)	0.6
Focused (external)	0.4
Focused (internal)	0.3
Minor factor	0.1

of actions and decisions directly under companies' control, unlike external factors. The provided overall framework about pros and cons (from a qualitative viewpoint) and the corresponding score (from a quantitative viewpoint) can represent per se a further variable for making decisions about the opportunities behind the design of PSS propositions.

Factors Operationalization

The calculations that follow are based on further assumptions, which are necessary to build a tentative quantitative equation capable of considering non-monetary aspects and other factors contextually.

In a company's perspective, SA should constitute a base for the estimation of a moderator or multiplier of expected economic benefits. In other terms, there is a relationship between sums calculated as in previous sub-paragraph and this multiplying coefficient, which was named Servitization Value Correction Coefficient (SVCC). In particular, the coefficient is meant to amplify or reduce economic values and/or indicators commonly employed in decision-making processes. Examples of indicators that can be used in combination with SVCC are those commonly involved in decision-making processes, like Net Present Value (NPV) or either Economic Value Added (EVA).

In case of a positive value of involved indicators, a profit is expected from a PSS introduction. In these circumstances, the presence of a positive (negative) value of CA results in boosting (moderating) the prospects of a positive outcome. The lowest value of SVCC should tend to 0 when disadvantages are largely predominant, and they can jeopardize the positive economic prospects of a PSS introduction. Of course, when $CA = SA + CC$ is equal to 0, the value of SVCC has to be 1—this means that equal advantages and disadvantages do not affect economic forecasts at all. To give a meaningful formulation in order to evaluate SVCC, we considered as a starting point the constraints given by limit values of $CA = SA + CC$ and expected corresponding values of SVCC, with EVA acting as a proxy for the implementation context (according to its sign). The idea behind the differentiation according to the value of EVA was a necessary point to provide an all-encompassing and meaningful employment of the proposed indicator SVCC in a decision-making process: in case of positive EVA, SVCC acts as an amplifier (in case of predominant advantages) or a limiter (in case of predominant disadvantages) of expected economic results;

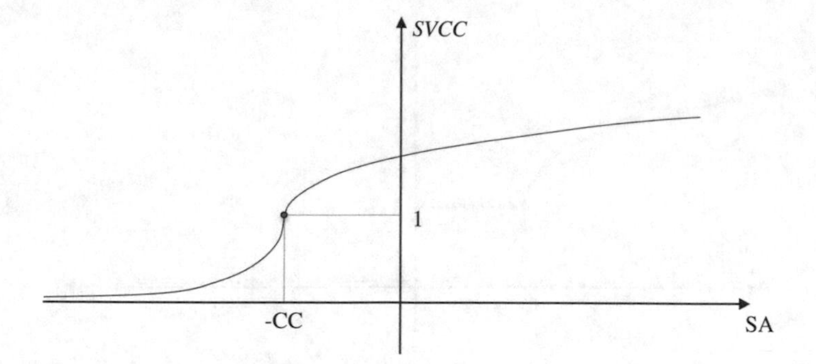

Fig. 4.11 Trend of SVCC+, EVA > 0

on the other hand, in case of negative EVA, SVCC is intended to minimize the loss (in case of predominant advantages) or to amplify the value destruction effect (with disadvantages overwhelming advantages).

- For EVA > 0, the following constraints must be considered:
  - For SA tending to $-\infty$, and for a given value of CC, SVCC tends to 0.
  - For SA + CC = 0, SVCC = 1.
- For EVA < 0:
  - For SA tending to $+\infty$, and for a given value of CC, SVCC tends to 0.
  - For SA + CC = 0, SVCC = 1.

In these constraints, SA was considered as the main independent variable since, as already explained, this variable represents the set of decisions under company's direct control (conversely from CC which synthetizes external factors that are, by definition, out of company's direct control).

According to considerations above, the formulation of SVCC is the following:

$$\mathrm{SVCC}^{+} = \begin{cases} e^{\mathrm{SA+CC}} & \mathrm{SA} + \mathrm{CC} \leq 0 \\ \ln(\mathrm{SA} + \mathrm{CC} + e) & \mathrm{SA} + \mathrm{CC} \geq 0 \end{cases}$$

$$\mathrm{SVCC}^{-} = \begin{cases} e^{-\mathrm{SA}-\mathrm{CC}} & \mathrm{SA} + \mathrm{CC} \geq 0 \\ \ln(-\mathrm{SA} - \mathrm{CC} + e) & \mathrm{SA} + \mathrm{CC} \leq 0 \end{cases}$$

With SVCC+ representing the case of positive EVA, while SVCC—is intended to be used in case of negative EVA. Figures 4.11 and 4.12 depict trend of SVCC functions.

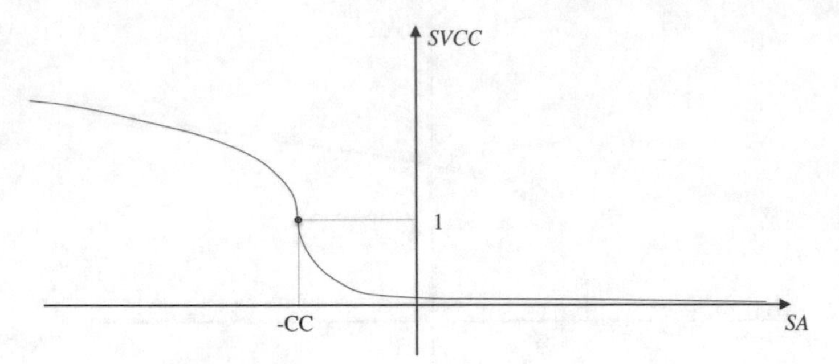

Fig. 4.12 Trend of SVCC−, EVA < 0

Seven Key Facts

- Translating product service system into a competitive strategy demands a considerable and careful effort in strategy formulation.
- When developing a servitization strategy, it is vital to focus on some distinctive strategic drivers while pursuing product/service differentiation and innovation.
- Companies should seize circular economy and sharing economy opportunities while at the same time securing their effort from replicability and imitability from competitors.
- The development of new market segments and building of customers' loyalty are key elements to ensure the sustainability of competitive advantage.
- The analysis and evaluation of risks connected to PSS is a fundamental part to build and secure a successful servitization strategy.
- The evaluation of competitive advantage sustainability must take into account revenue flows, cost structure and the added value of PSS.
- Non-monetary factors play a crucial role in affecting the overall success of a PSS-based offering and must be taken into account in decision-making efforts and as a side aspect of economic evaluations.

References

R. Amit, P.J.H. Schoemaker, Strategic assets and organizational rent. Strateg. Manag. J. **14**, 33–46 (1993)

A. Annarelli, C. Battistella, F. Nonino, Product service system: a conceptual framework from a systematic review. J. Clean. Prod. **139**, 1011–1032 (2016)

W.B. Arthur, Positive feedbacks in the economy. Sci. Am. **262**, 92–99 (1990)

A. Azarenko, R. Roy, E. Shehab, A. Tiwari, Technical product-service systems: some implications for the machine tool industry. J. Manuf. Technol. Manage. **20**(5), 700–722 (2009)

A. Azevedo, Innovative costing system framework in industrial product-service system environment. Proc. Manufact. **4**, 224–230 (2015)

J.B. Barney, *Gaining and Sustaining Competitive Advantage* (Prentice Hall, Upper Saddle River, NJ, 2002)

S. Bonnemeier, F. Burianek, R. Reichwald, Revenue models for integrated customer solutions: concept and organizational implementation. J. Revenue Pricing Manag. **9**(3), 228–238 (2010)

R. Botsman, R. Rogers, *What's Mine is Yours, the Rise of Collaborative Consumption* (Harper-Collins, New York, NY, 2010)

K. Chirumalla, A. Bertoni, A. Parida, C. Johansson, M. Bertoni, Performance measurement framework for product-service systems development: a balanced scorecard approach. Int. J. Technol. Intell. Planning **9**(2), 146–164 (2013)

C.-J. Chou, C.-W. Chen, C. Conley, An approach to assess sustainable product-service systems. J. Clean. Prod. **86**, 277–284 (2015)

A. Davies, Moving base into high-value integrated solutions: a value stream approach. Ind. Corp. Change **13**(5), 727–756 (2004)

R. De Coster, A collaborative approach to forecasting product-service systems (PSS). Int. J. Adv. Manuf. Technol. **52**, 1251–1260 (2011)

A. Dimache, T. Roche, A decision methodology to support servitization of manufacturing. Int. J. Oper. Prod. Manage. **33**(11–12), 1435–1457 (2013)

U. Gelbmann, B. Hammerl, Integrative re-use systems as innovative business models for devising sustainable product-service systems. J. Clean. Prod. **97**, 50–60 (2015)

C. Grönroos, A service perspective on business relationships: the value creation, interaction and marketing interface. Ind. Mark. Manage. **40**(2), 240–247 (2011)

T. Hill, *Manufacturing Strategy: Text and Cases* (MacMillan Business, 1994)

H.A. Hu, S.H. Chen, C.W. Hsu, C. Wang, C.L. Wu, Development of sustainability evaluation model for implementing product service systems. Int. J. Environ. Sci. Technol. **9**, 343–354 (2012)

W.Q. Judge, T.J. Douglas, Performance implication of incorporating natural environmental issues into the strategic planning process: an empirical assessment. J. Manage. Stud. **35**(2), 241–262 (1998)

D. Kindström, Towards a service-based business model—key aspects for future competitive advantage. Eur. Manag. J. **28**(6), 479–490 (2010)

D. Kindström, C. Kowalkowski, Service innovation in productcentric firms: a multidimensional business model perspective. J. Bus. Ind. Mark. **29**(2), 96–111 (2014)

P. Kotler, *Marketing Management*, The Millennium Edn. (Person Prentice Hall, Upper Saddle River, 2000)

M.E. Kreye, L.B. Newnes, Y.M. Goh, Uncertainty in competitive bidding—a framework for product–service systems. Prod. Plann. Control **25**(6), 462–477 (2014)

S.J. Liebowitz, S.E. Margolis, Path dependence, lock-in, and history. J. Law Econ. Organ. **11**(1), 205–226 (1995)

E. Manzini, C. Vezzoli, A strategic design approach to develop sustainable product service systems: example taken from the 'environmentally friendly innovation' Italian prize. J. Clean. Prod. **11**, 851–857 (2003)

V. Martinez, M. Bastl, J. Kingston, S. Evans, Challenges in transforming manufacturing organisations into product-service providers. J. Manufact. Technol. Manage. **21**(4), 449–469 (2010)

P. Matthyssens, K. Vandenbempt, Service addition as businessmarket strategy: identification of transition trajectories. J. Serv. Manage. **21**(5), 693–714 (2010)

H. Meier, R. Roy, G. Seliger, Industrial product-service system—IPS2. CIRP Ann. Manuf. Technol. **59**, 607–627 (2010)

G.A. Michaelson, Niche marketing in the trenches: Marketing Communications. **13**(6), 19–24 (1988)

O. Mont, Clarifying the concept of product-service system. J. Clean. Prod. **10**, 237–245 (2002)

O. Mont, Product–service system: panacea or myth? Doctoral thesis. Retrieved from the National Library of Sweden Database. 91-88902-33-1 (2004)

T. Nagle, J. Hogan, *The Strategy and Tactics of Pricing—A Guide to Growing More Profitably*, 4th edn. (Pearson Education, Upper Saddle River, NJ, 2006)

A. Neely, Exploring the financial consequences of the servitization of manufacturing. Oper. Manage. Res. **1**, 103–118 (2009)

R. Nelson, S. Winter, *An Evolutionary Theory of Economic Change* (Harvard University Press, Cambridge, MA, 1982)

S. Nenonen, K. Storbacka, Business model design: conceptualizing networked value co-creation. Int. J. Qual. Serv. Sci. **2**(1), 43–59 (2010)

N. Nishino, S. Wang, N. Tsuji, K. Kageyama, K. Ueda, Categorization and mechanism of platform-type product-service systems in manufacturing. CIRP Ann. Manuf. Technol. **61**, 391–394 (2012)

A.F. Payne, K. Storbacka, P. Frow, Managing the co-creation of value. J. Acad. Mark. Sci. **36**(1), 83–96 (2008)

M.E. Porter, *The Competitive Advantage: Creating and Sustaining Superior Performance* (Free Press, NY, 1985)

M. Qu, S. Yu, D. Chen, J. Chu, B. Tian, State-of-the-art of design, evaluation, and operation methodologies in product service systems. Comput. Ind. **77**, 1–14 (2016)

C. Raddats, C. Easingwood, Service growth options for B2B product-centric businesses. Ind. Mark. Manage. **39**(8), 1334–1345 (2010)

M. Rapaccini, F. Visintin, Devising hybrid solutions: an exploratory framework. Prod. Plann. Control **26**(8), 654–672 (2015)

W. Reim, V. Parida, D. Örtqvist, Product-service systems (PSS) business models and tactics—a systematic literature review. J. Clean. Prod. **97**, 61–75 (2015)

W. Reim, V. Parida, D. Rönnberg-Sjödin, Risk management for product-service system operation. Int. J. Oper. Prod. Manage. **36**(6), 665–686 (2016)

W. Reim, V. Parida, D. Rönnberg-Sjödin, Managing risks for product-service systems provision: introducing a practical decision tool for risk management, in *Practices and Tools for Servitization—Managing Service Transition*, ed. by M. Kohtamaki et al. (Palgrave MacMillan, Cham, 2018), pp. 309–321

M. Rese, M. Karger, W.C. Strotmann, The dynamics of industrial product service systems (IPS2)—using the net present value approach and real options approach to improve life cycle management. CIRP J. Manuf. Sci. Technol. **1**, 279–286 (2009)

O. Rexfelt, V. Hiort af Ornas, Consumer acceptance of product-service systems. Designing for relative advantages and uncertainty reductions. J. Manuf. Technol. Manage. **20**(5), 674–699 (2009)

A. Richter, T. Sadek, M. Steven, Flexibility in industrial product-service systems and use-oriented business models. CIRP J. Manuf. Sci. Technol. **3**, 128–134 (2010)

M. Sawhney, Going beyond the product: defining, designing, and delivering customer solutions, in *The Service-dominant Logic of Marketing. Dialog, Debate, and Directions*, ed. by R. Lusch, S. Vargo (M.E. Sharpe, Armonk, NY, 2006), pp. 365–380

M. Spring, L. Araujo, Service, services and products: rethinking operations strategy. Int. J. Oper. Prod. Manage. **29**(5), 444–467 (2009)

K. Storbacka, A solution business model: capabilities and management practices for integrated solutions. Ind. Mark. Manage. **40**(5), 699–711 (2011)

H. Sun, Z. Wang, Y. Zhang, Z. Chang, R. Mo, Y. Liu, Evaluation method of product-service performance. Int. J. Comput. Int. Manuf. **25**(2), 150–157 (2012)

A.R. Tan, D. Matzen, T. McAloone, S. Evans, Strategies for designing and developing services for manufacturing firms. CIRP J. Manuf. Sci. Technol. **3**(2), 90–97 (2010)

D.J. Teece, G. Pisano, A. Shuen, Dynamic capabilities and strategic management. Strateg. Manag. J. **18**(7), 509–533 (1997)

A. Tukker, Eight types of product-service system: eight ways to sustainability? Experience from SusProNet. Bus. Strategy Environ. **13**, 246–260 (2004)

A. Tukker, U. Tischner, Product-service as a research field: past, present and future. Reflection from a decade of research. J. Clean. Prod. **14**, 1552–1556 (2006)

J. Turner, S. Simister, Project contract management and a theory of organization. Int. J. Project Manage. **19**(8), 457–464 (2001)

J. Van Ostaeyen, A. Van Horenbeek, L. Pintelon, J.R. Duflou, A refined typology of product–service systems based on functional hierarchy modeling. J. Clean. Prod. **51**, 261–276 (2013)

S. Witjes, R. Lozano, Towards a more circular economy: proposing a framework linking sustainable public procurement and sustainable business models. Resour. Conserv. Recycl. **112**, 37–44 (2016)

Chapter 5
Translating PSS Strategy into Operations

The chapter deals with the importance of an effective translation of strategy (analysed in Chap. 4) into operations strategy and operations management.

After an introductory part on the importance of operations and service strategies, the key elements for the effective "translation" are presented and discussed.

A series of key activities is essential for the design of PSS in line with strategic directions, while key resources and partners ensure an optimal implementation of PSS.

5.1 The New Role of Operations and Service Strategy

As highlighted in previous chapters, the shift towards servitization and the adoption of PSS have a considerable impact on company's overall structure. The reconfiguration of the whole business model impacts as well on the entire set of operations. Indeed, once the business strategy has been redefined accordingly, managers must reconfigure to the operations strategy and then consequently the operations.

The framework in Fig. 5.1 represents the process of PSS configuration: it consists of seven elements divided into three groups (Aurich et al. 2009).

The first group considers all elements that concern prerequisites of PSS development: physical components, service components and the product life-cycle perspective. In the second group, the two elements are the influence of services and life-cycle perspective over the product and the physical components of the offering. Here there is the shift of attention from business strategy to operations strategy and first critical elements. The third group, concerning more technical-specific aspects of PSS configuration, contains elements that focus on the process of actual PSS design, representing technical issues and service configuration issues that are the "final" elements in determining a PSS tailored for customer-specific needs.

Looking at Fig. 5.1, the key role of translating PSS into operations strategy is represented by the elements of the second and third groups.

A. Annarelli et al., *The Road to Servitization*,
https://doi.org/10.1007/978-3-030-12251-5_5

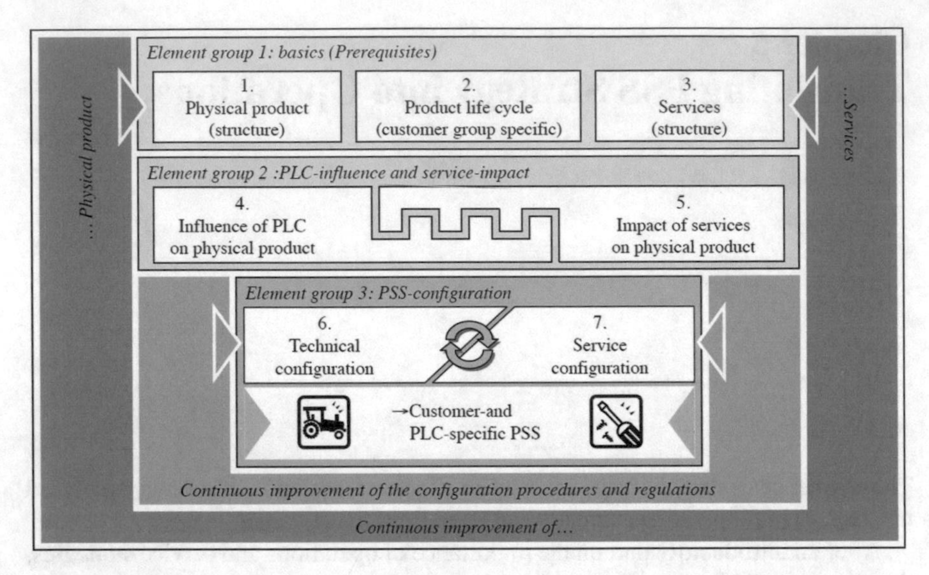

Fig. 5.1 Framework for PSS configuration (Aurich et al. 2009)

Furthermore, in this critical step, companies should also keep in mind the variety of elements being involved in a PSS, and their different origins. For instance, PSS has many elements in common with the Service Engineering field, considering that the service component is considered to be one of the most critical elements when servitizing: indeed, practice-based evidence suggests that the majority of companies, undergoing this transformation, start from a product-focused offering, where the standard offering focuses only on stand-alone sales.

5.2 Key Activities for Product Service System Design

PSS providers must focus on their customers' key activities, rather than on the physical characteristics of their product. With PSS, the (customer-supplier) processes and operating activities are integrated and, if a specific function is assigned to a product, essential activities are linked to it before, during and after the actual use phase (Cook et al. 2006). During the use phase, in fact, the producer can monitor the performance of the product and plan maintenance interventions (Schuh et al. 2009). The integration between operations and related activities must be managed cautiously both from a tactical and a strategic point of view since a new orientation to support the PSS is needed. Order making, storage, cost control, installation, use, maintenance and troubleshooting can be typical activities incorporated into a PSS (Grönroos 2011). The identification activity highlights the processes critical for the success of the development and service delivery (Lay et al. 2009; Kindström and Kowalkowski 2014). Switching to PSS logic can push companies to outsource activities that were previously

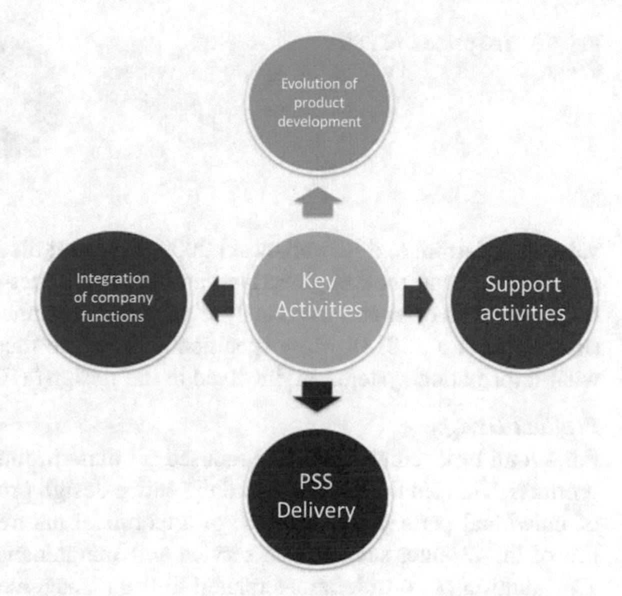

Fig. 5.2 Identifying key activities for PSS design

developed within company boundaries (Storbacka 2011; Dimache and Roche 2013). The evolution towards a logical service may require a company to redesign its internal organization in order to implement new service-oriented activities by identifying key activities that must in the transition from traditional contexts to those typical of PSS (Fig. 5.2).

5.2.1 *Product and Service Design*

The first activity concerns the development and design phase of the solutions that the company intends to offer. In the PSS business model, to meet product and service specifications, special emphasis is given to the alignment between physical characteristics of the product with the service (Reim et al. 2015). Furthermore, the company is usually responsible for all costs incurred during the life cycle of the product. This leads us to rethink the design phase so as to minimize the overall costs incurred during the life cycle of the product (life extension, reduction of operational costs), to make the asset easy to maintain (Azarenko et al. 2009) and insert parts which can be reused at the end of the product life cycle (Tukker 2004). Many properties of the product in terms of maintainability, ease in updating and reutilization are identified and developed in the designing phase in order to facilitate the administration of the product related to the service in question and increase the value created in the new business model (Sundin and Bras 2005). So, the quality of service is linked to the ideation/innovation of service components that can improve the PSS offer in order to better interpret the client's requests and make the creation of the most performing

Fig. 5.3 The process of PSS design

value (Kindström and Kowalkowski 2009; Pawar et al. 2009). The development of new services and their engineering can help a product-centric activity to successfully extend its offer and then catalyse a drive for change towards servitization logic (Rapaccini et al. 2013). More specifically, how are these activities structured and what information systems are involved in the design? (Fig. 5.3)

Product Design
For a long time, companies have focused on mass production and physical sales of products. This led the engineers and the entire design process to focus on the design of individual parts. Some services of a technical nature were added to extend the life of the product such as tele service and maintenance (Maussang et al. 2009). The addition of considerations related to the management of the life cycle has led to greater attention to the physical availability of the product. The weaknesses were dealt with during the design phase by inserting technical features that in the future would be better interfaced with the maintenance activities.

Service Design
Usually, the design of the services was entrusted to the marketing operators (Maussang et al. 2009) unlike the physical products that were usually the prerogative of the engineering component. A study conducted by Tomiyama (2001) led to a methodology for developing services. The methodology has three basic steps as follows:

- The flow model: Allows the developer to map all the agents involved in the delivery of the service and its interactions.
- The scope model: Focuses on the parameters that are impacted by the service.
- The scenario model: This model displays the set of customer parameters and related channels.

PSS Design
The phase related to the design and development of PSS is more complex and strategic than the isolated design of a product or service disconnected from each other or not related to other components of the same category. In this phase, we weigh aspects from different sectors, from marketing to the production context, to the necessary involvement of the client for co-creation. With the appearance of the PSS solutions, the market demands have changed and, in order to align the new prerequisites, special emphasis is given to the alignment between the physical characteristics of the product and the characteristics of the service (Reim et al. 2015). To increase the value of the offer, some aspects such as ease of maintenance, updating and use cannot be underexamined. Everything can be summarized in two essential prerogatives that must be considered during the design and development phase of the PSS: aspects

related to functionality and customization. The former considers the need for the various components to be designed to subsequently incorporate a further one in a way that is positively perceived by the customer. For product-oriented solutions, this implies that for the products there is a maintainability in the case of contracts that provide for maintenance, or that the parts are easily recoverable in the case of contracts that provide for disposal (Reim et al. 2015). In use-oriented contracts, the provider is responsible for the use of the asset, so in addition to guarantee a simple maintainability, the design must guarantee the strength and durability of the components. As regards the result-oriented context, the concept of functionality is reinforced by the need to create a flexible solution through the combination of various components and services (Reim et al. 2015). The second aspect concerns personalization that is the need to adapt the solution to each individual customer. In product-oriented and use-oriented contexts, personalization is relatively low and takes on a more important meaning for result-oriented solutions, given that the service must perfectly blend with the client's production and organizational system, which is unique for every case. In this case, the design of the services associated with the package of solutions offered must be carried out carefully and involving the customer (Reim et al. 2015).

Given these two aspects, it is possible to list some techniques used for the development and design of PSS (Table 5.1).

These methodologies, summarized in the table, can be seen more in detail as given below:

Table 5.1 Design methods in PSS context (Vasantha et al. 2011)

Method	First granted US patent (date)
Service CAD	Method for designing business models that increase eco-efficiency
Service model Service explorer	It focuses on the design of the service to offer products with a high added value
Integrated design process	It focuses on potential interactions between product and service and then decides how to redefine actual design activities
Fast track total care design process	Integrates product and service features
PSS design	Provides guidance on adding value during the design phase
Heterogeneous concept modelling	Model-based approach
Dimension of PSS design	Description of the new concepts inherent in the PSS
Design for integrated solution	A methodological development based on the creation of new tools

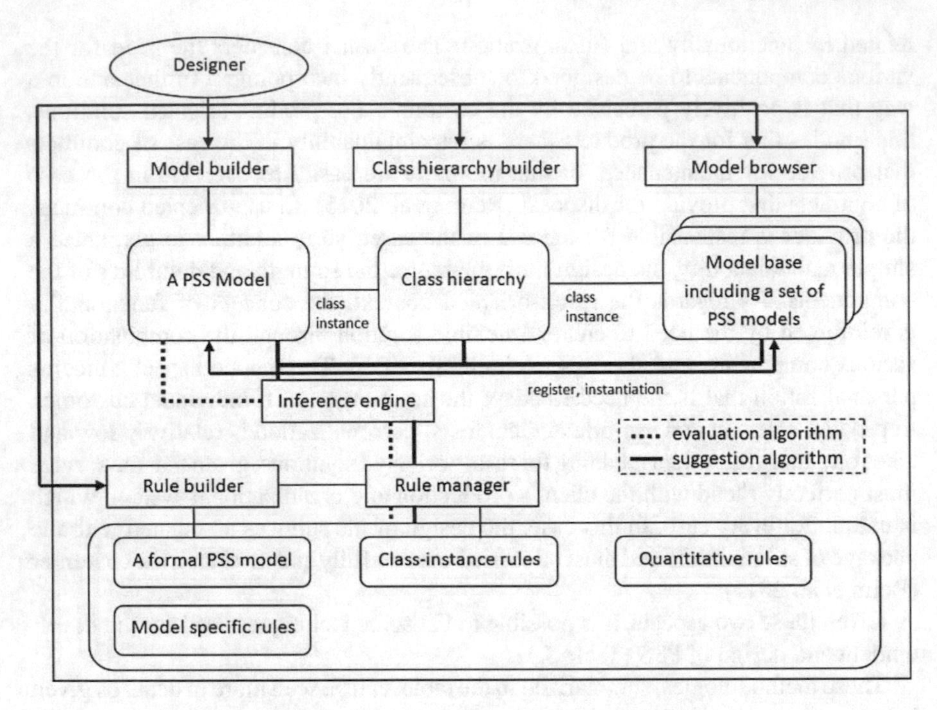

Fig. 5.4 The Service CAD architecture (Komoto and Tomiyama 2008)

- Service CAD. CAD service systems help designers to generate a design structure of the PSS. The structure helps designers to detect problems and allows them to conceptualize and suggest alternatives. According to this method, designers define activities to meet specific performance and quality prerequisites. The variables present in the service CAD are the operating context, the provider, the customer, the channel, the activity, the intentions of use, the foreseen, the quality and the added value. There is an extension of the system called ISCL (integrating Service CAD with a life-cycle simulator) that simulates the life cycle of the product in which there are probabilistic descriptions linked to the consequences of the use activities. This method can be used as a representation in which the designer can construct a model of PSS in different ways. Given the uniqueness of the solutions, it is difficult to categorize the variables seen previously. In addition it is difficult to identify activities that may overlap and synchronize the activities of the product and the process from a temporal point of view (Fig. 5.4).
- Service explorer. According to this approach, the product and service are designed simultaneously during the early stages of PSS development. The goal is to maximize the value for the customer, considering synergies, alternatives and complementarity. The method proposes a uniquely schematized representation of the human and physical aspects that emerge during the service activity and is composed of three main phases: identification of the value for the client, planning of the contents of the service and planning of the service activities. With this method,

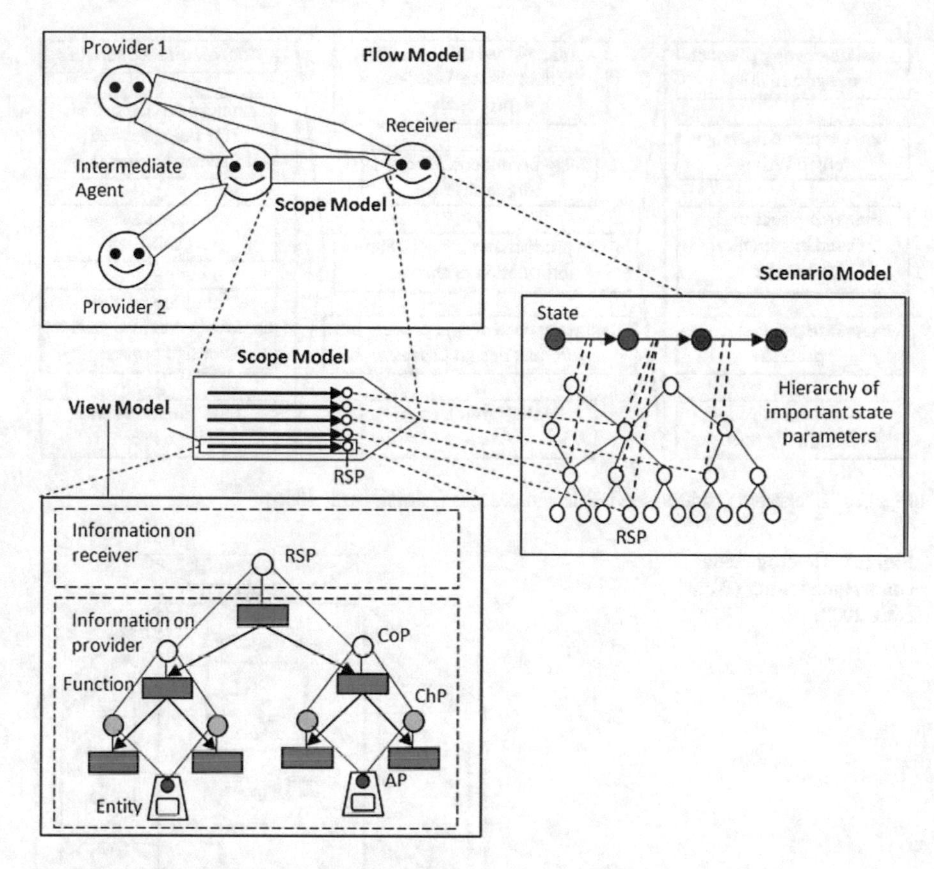

Fig. 5.5 Relationship between flow, scope and scenario models (Sakao et al. 2009)

there is an evaluation of the processes through the Quality Function Deployment (QFD) tool. From the graphical point of view, it is possible to identify three blocks: the flow model (who), the scope model (what) and the model scenario (why) and the view model (how) (Fig. 5.5).

- Integrated design process. This method introduces a process for the design of technical services that match the product according to a logic for modules. Services and products are developed in parallel and then integrated into the reality of the PSS (Fig. 5.6).
- Heterogeneous concept modelling. This approach allows the multidisciplinary combination of elements on different levels of abstraction from different development perspectives. This leads to the need for simulations to determine the behaviour of processes and components. The approach is implemented as a software prototype and is developed using three types of elements: system elements, disturbance elements and context elements. There are five elements that characterize future PSS solutions: specificity, main transformation, level of integration, predisposition to partial substitutions and connectivity (Fig. 5.7).

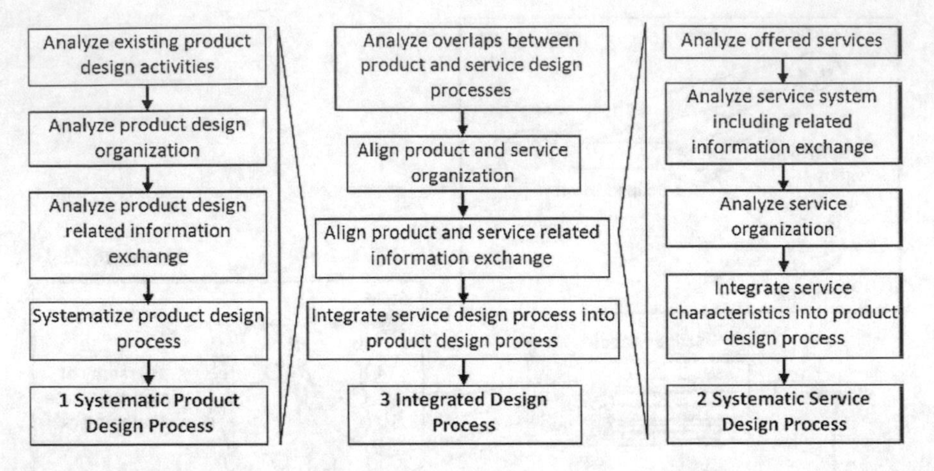

Fig. 5.6 Integrated product–service design process (Aurich et al. 2006)

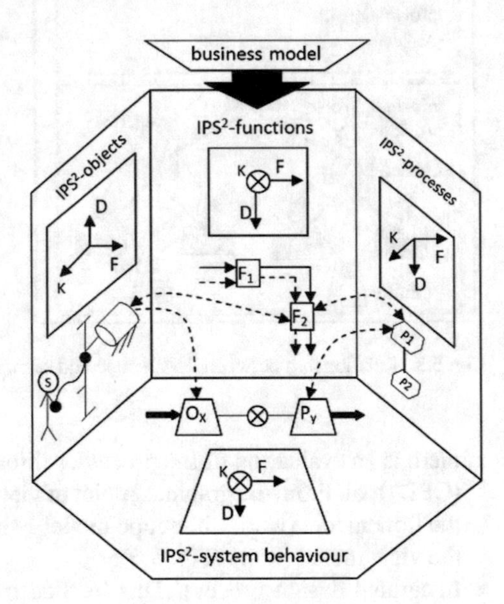

Fig. 5.7 Heterogeneous concept modelling (Welp et al. 2008)

- PSS design. According to this approach, the attention of the designers is focused not only on the physical aspects of the product, but above all on the whole system dictating the physical part and the service necessary to develop a successful PSS. An operational scenario is used to go into more detail once the main elements (physical and service components) have been identified. It is good practice to simulate an operational scenario for each phase of the life cycle (Fig. 5.8).
- Fast track design. According to this method, there are two aspects to consider: architecture and business. The architecture consists of hardware and support services, while the business aspect encompasses all the nuances related to markets, risks, partnerships, contracts, sales and distribution of the solutions. These two

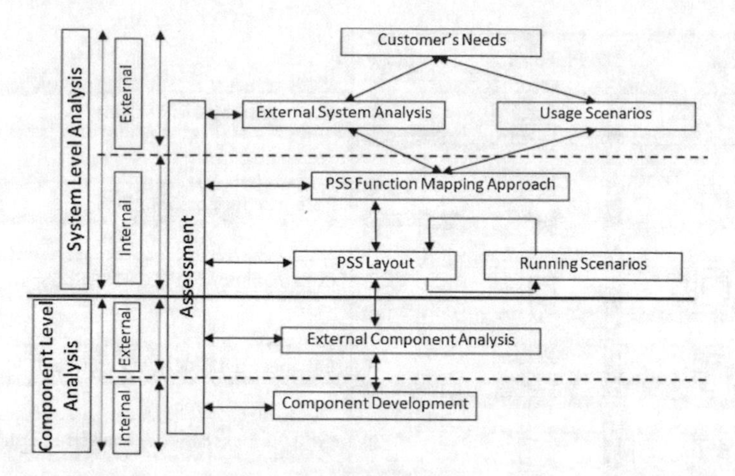

Fig. 5.8 PSS design (Maussang et al. 2009)

categories describe the different combinations of hardware and services: new hardware, products adapted from previous ones, new support services or adapted support services. The aim is to choose the most suitable combination of products and services to create the best solution for all the parties involved. Operatively, this involves the integration between the communication of customer needs and the conceptualization of product features. To perform this step correctly, the company needs an IT tool that integrates service design, simulation, hardware architecture and a system to calculate support service costs. The overall analysis makes it possible to identify critical elements, costs/benefits and required resources. The implementation of this methodology helps both customer and supplier, reducing the complexity of the process by simplifying the decision-making phase and the analysis of alternatives (Fig. 5.9).

- Design process for development of service. This model consists more of a theoretical approach than a precise technical practice and consists of a rearrangement of the logic design sequences in the service context. Graphically, the method can be expressed by dividing space into two dimensions: the space dedicated to problems and the space dedicated to design solutions related to the problems identified. Typical elements present in the part related to the problems concern the market analysis, the hypotheses of use and the simulations (Fig. 5.10).
- PSS process design. The methodology outlined here requires four categories to be set spatially, which are the value proposition, the life cycle of the product, the network of actors and the activity modelling cycle. These elements cover all the essential concepts of the PSS. An accurate analysis of these dimensions is a good way to understand how products and systems work and allows to underline those components in which the four dimensions must be aligned. It is important to note that a change of one of these four dimensions influences the others and the designers must be sure that each dimension of a new PSS supports the others. According to this approach, the design of a PSS relates to the management, the

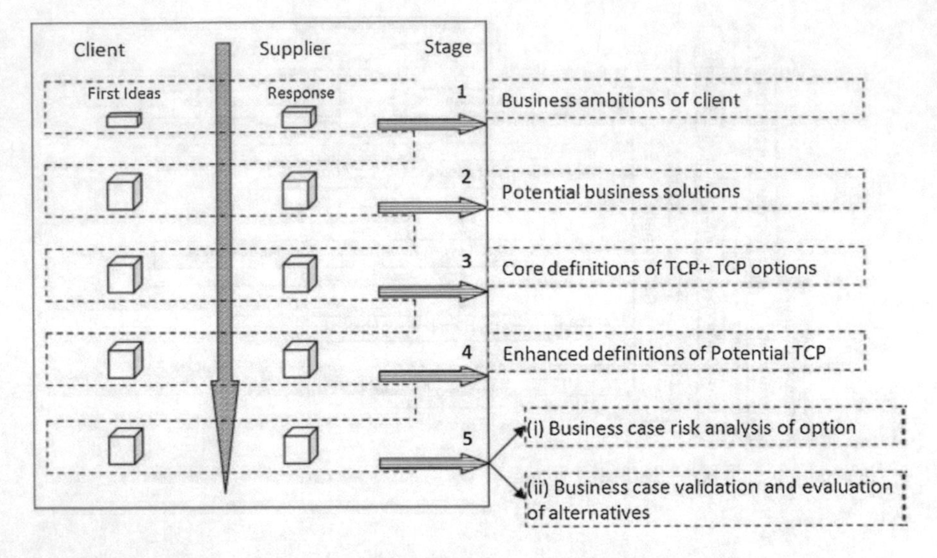

Fig. 5.9 Fast track design process (Alonso-Rasgado et al. 2004)

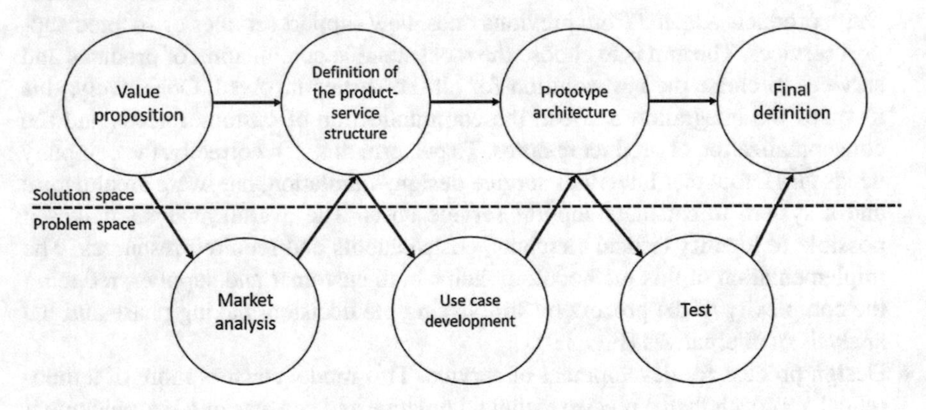

Fig. 5.10 Schematic view of service development process (Morelli 2002)

organizational structure, the coordination and the integration of the development activities that are not covered in the four dimensions (Fig. 5.11).

5.2.2 *Product and Service Configuration Support*

Support for a new type of solution consists of a series of activities aimed at making the customer understand the new potential offered. If it is true, the PSS consists of a new offer with specific product and service configurations that are set up to create value for the individual customer and it is equally true that the offer of PSS is complex

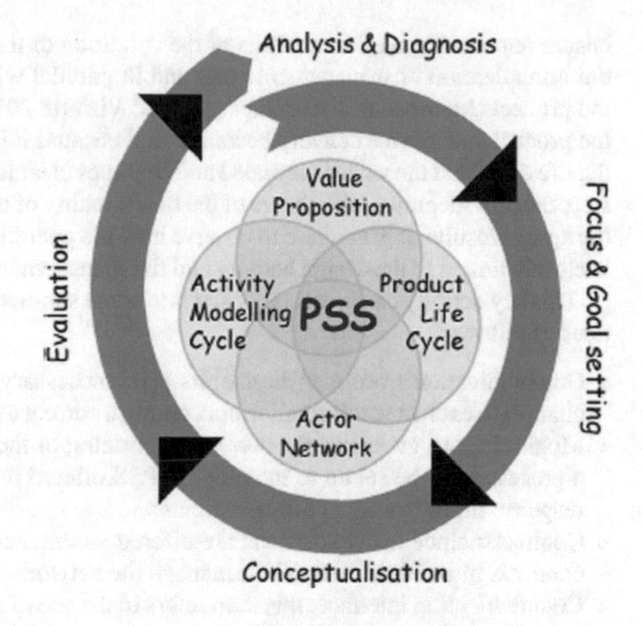

Fig. 5.11 Phases of PSS design process (Tan et al. 2009)

and that focusing on value-driven logic becomes central when the company needs to show the potential customer its new offer. In this regard, there are specific strategies and methods (e.g. total cost of ownership, service-level agreement) to help the client appreciate the improvements and benefits of the PSS. Given the thickness of the technological level contained in the PSS offer, the front-end must be adequately trained in the conviction and in the transmission of the value to the customer (Kindström and Kowalkowski 2009, Kindström et al. 2015). It is therefore necessary to train sales forces from a technical point of view to transmit the improvements and advantages to the customer in a clear and simple way. During the sales phase the total cost of ownership or the life cycle costing evaluation can be useful. These applications are still pioneering and addressing few possible product and service configurations (Bonetti et al. 2016).

5.2.3 *Product and Service Delivery*

In PSS logic, service delivery planning and related operational activities must be carefully managed to ensure that integrated systems are accessed efficiently (Storbacka 2011). To allow rapid acceleration in the delivery of operational activities and extensive incorporation with the customer the process must be constantly monitored carefully. It is necessary to verify and report to both customers and internally whether the planned value has actually been created and to document the delivery (Kindström and Kowalkowski 2009). Providing a service does not mean guaranteeing only spare parts, operational information and routine maintenance, but it is a commitment to ensure remote diagnostics, analysis of the condition of the product that is crucial in the minimization of management costs and in parallel with the value generated by the product (Meier et al. 2010; Rapaccini and Visintin 2015). The activity related to the product and service delivery becomes vital because it is precisely in this phase of the life cycle that the value is revealed and becomes clear to the customer. During this step, the provider must take charge of the functionality of the solutions and guarantee the agreed results. It is intuitive to observe how the operation of the solutions offered is closely linked to the design activity and the management of the product life cycle.

This key activity can be decomposed into some sub-activities or managerial practices as follows:

- Communication process: in the use phase, it is necessary to set up a communication channel to exchange information concerning a correct use of the solution provided.
- Monitoring: to guarantee the correct functioning of the solution to the customer, a process must be set up to monitor the PSS offered to the client to optimize the response times in case of adverse events.
- Contracts: since many solutions are offered simultaneously to several partners, contracts must be defined to best manage the network.
- Communication interface: this term refers to the use of extranet platforms to guarantee immediate access to information from all network actors (Fig. 5.12).

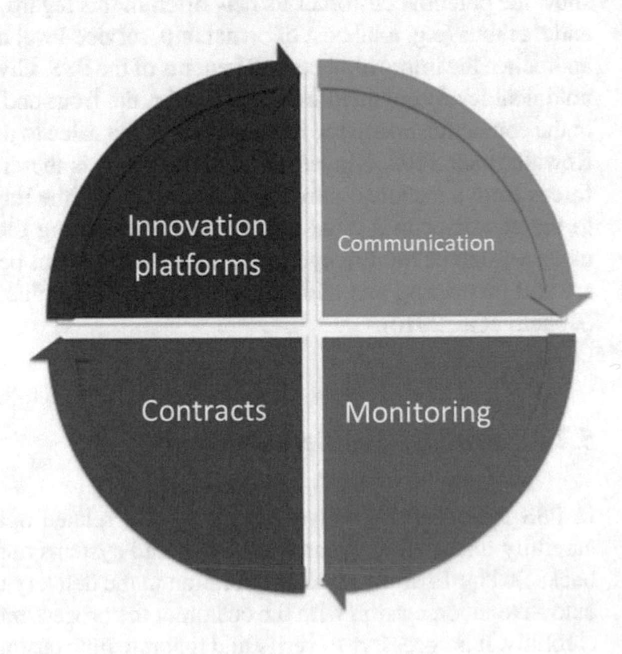

Fig. 5.12 PSS delivery process

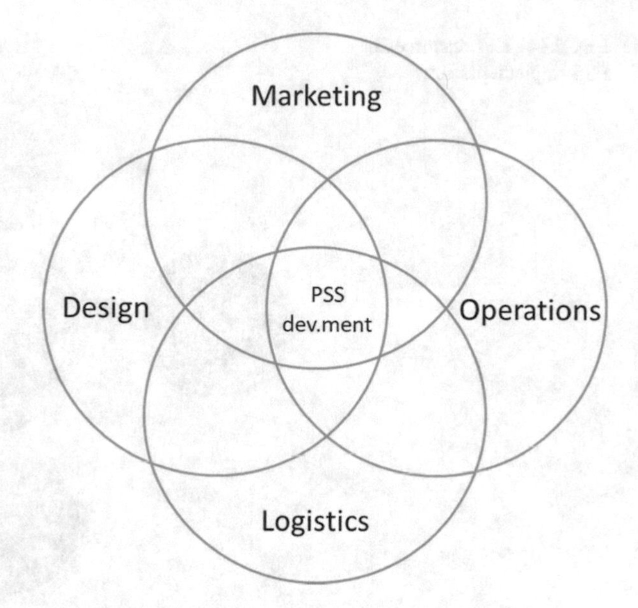

Fig. 5.13 Functional integration in PSS context

5.2.4 *Functional Integration*

The transition to PSS contexts requires formal processes and mechanisms in order to realize intra-functional activities (Kindström et al. 2015) and achieve an integration between the different company functions (Nordin and Kowalkowski 2010; Storbacka et al. 2013). Consider how in the design phase the marketing units and the technical offices are required to work in harmony in order to develop solutions that are congruent with the needs of the customer. More in detail, the PSS BMs require a collaboration between the implementation of the service and R&D (Kindström and Kowalkowski 2014). As we have seen above, the reality of PSS solutions is not absolutely trivial or simplistic and its management must be set considering the whole package as a reality in its own right and must be managed in its entirety. With the increasing complexity of the offer, there is the need to coordinate every step of the product life cycle both before and after the sale, guaranteeing the customer a single point of contact with the supplier company (Kindström et al. 2015). In other words, the PSS BMs need collaborative management and the evaluations on the quality of the work have to be considered in this new intra-functional nature (Storbacka 2011) (Fig. 5.13).

5.3 Key Resources in PSS Implementation

Key activities can be defined as the set of processes that are at the core of the success of the development and delivery of the solution offered. To undertake the

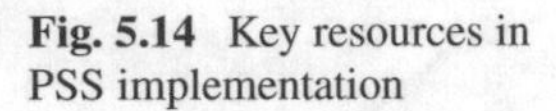
Fig. 5.14 Key resources in PSS implementation

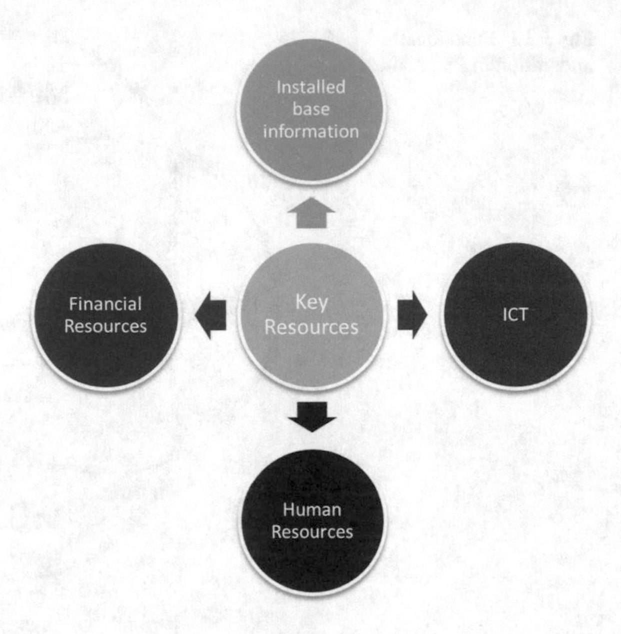

implementation of the PSS systems companies need to see what new key resources must be used to successfully address this challenge. If, as we have seen so far, the relationships between the various actors play a very important role, it is not surprising to know that PSS providers must invest a lot of resources in human capital (Tan and Mcaloone 2006). The quality of the human capital must be unexceptionable. In this sense, new skills must be developed in order to create a competent and reassuring interface with customers focusing both on training courses for staff in force and with new hires (Cook et al. 2006). Another key resource that must be provided at the point of contact between the customer and the supplier is almost never necessarily physical, and often consists of ICTs in which information exchange takes place. It is therefore logical that it is necessary to invest in an infrastructure capable of facilitating the relationship between customer and supplier. In companies based on PSS logic, some activities, previously carried out internally, can be outsourced, thus requiring resources that go beyond the boundaries of the company (Dimache and Roche 2013). As seen previously, service innovation may require an organizational change (Kindström and Kowalkowski 2009) in order to develop new activities related to the service offered (Cavalieri and Pezzotta 2012) (Fig. 5.14).

5.3.1 ICT and Monitoring Technologies

ICT and digital technologies are at the basis of the implementation of PSS (Becker et al. 2013; Ardolino et al. 2018). ICT systems allow to share information and analyses extracted from data collected in different functions (Storbacka 2011). The classic management systems (from ERP to PDM) must be integrated in a fluid and substantial way, in order to coherently support applications aimed at supporting collaboration along the supply chain (Neff et al. 2014). The use of remote control technologies is essential in PSS logic to ensure supervision, maintenance and upgrades.

Currently, the effectiveness of PSS solutions is based on a large number of ICT tools, which support and standardize processes. Understanding the relationship between ICT and PSS is fundamental to understanding the impact of ICT on service quality. It has been shown that investment in ICT increases the economic performance of companies that adopt this type of system, as well as their financial and organizational well-being. So, ICT solutions can be seen as one of the key tools an organization must equip itself in order to ensure added value to its solutions. The aspects on which ICT tools impact are essentially two: product offering and operational processes (Belvedere et al. 2013).

Product offering. The development of information and communication technologies has led to obvious changes in the offer. Primarily, ICTs have led to a marked increase in the efficiency and effectiveness of the process of developing new PSS solutions making their value proposition more attractive. Secondarily, they have allowed the development of completely new solutions, reinforcing the company's competitiveness (Belvedere et al. 2013).

Concerning ICT and product development, new communication technologies have made possible new configurations of their own network, facilitating innovations and developing the capacity for simultaneity in designing between distant actors. The parallel growth of decision-making instruments strongly influenced the New Product Development process. This is the typical case of Product Lifecycle Management (PLM) that allows designers to store and share documents and related information, so as to reduce errors and redundancies in process innovation. Further help from this type of technology comes from their ability to capture the "customer's voice", for example, through online questionnaires. In the final analysis, we must also consider the use of Internet tools to distribute products or music (Fisher 2004).

As regards ICT and servitization, competitive context is even more characterized by new business models based on mixed product–service solutions. In the product use context the physical product is still sold to the customer and the value proposition is enriched by the service component; according to Simmons (2001), this strategy can be implemented thanks to the ICT tools that allow an exchange of information in real time. Thanks to these tools, it is therefore possible to observe the wear levels of the components or their actual availability. However, if in the previous case the positive impact of ICTs on the development and design phase was incontrovertible, in the case of development to servitization, ICTs positively impact only when the operational processes are redesigned around their operational needs (Belvedere et al. 2013).

Operational processes. There is often the problem of how to set up operational processes in companies that want to offer a set of products and services. This type of problem is central, especially in companies that intend to offer after-sales services such as the supply of spare parts and maintenance services. Such services need an adequate structuring of relatively unorganized processes. Using information technology tools can be a method for developing and redesigning processes (Belvedere et al. 2013). The beneficial effect is double and can be summarized by counting the standardization of operating processes and the increase in their response capacity.

About process standardization, one of the main aims of the use of information platforms is that linked to the automation of processes. This is made possible by the operational interpenetration between systems and processes. The standardization of processes and the unification of information platforms allow, in the case of sales phases, to reduce errors or mitigate the damage caused by untrained personnel (Buttle et al. 2006). In addition to standardization, there is also a greater diffusion of best practices within the organization (Belvedere et al. 2013). These two aspects, when combined together, lead to greater productivity and an increase in the reactivity of the processes. Although not relevant to standardization, other operational advantages need to be mentioned. In addition to the benefits due to scheduling tools, it should be noted that in the design phase, it is possible to use ICT tools such as CAD (Computer-Aided Design) and CAQ (Computer-Aided Quality assurance) that allows to reduce costs and lead times.

About process reactivity, on the other hand, reactivity can be observed from three points of view: volumes, products and processes (Holweg 2005). The first dimension refers to the ability to modify the volumes produced according to the demand peaks. The second is the ability to add new products to the production line and the third refers to the company's ability to produce and quickly deliver its products. So, a developed control of data coming from the production phases, thanks to ICTs, makes it possible to overcome the numerous inefficiencies in which it is possible to run throughout the entire supply chain, or to prevent the Forrester effect[1]. In fact, it is possible to immediately inform the customer of any delays or immediately consider exceptional measures to schedule the activities and guarantee deliveries within the pre-established deadlines (Fig. 5.15).

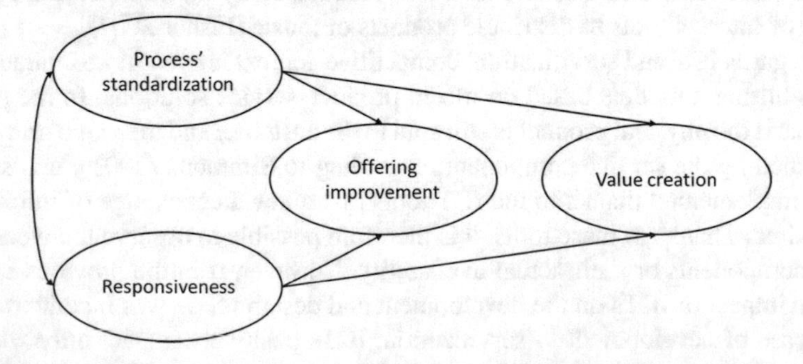

Fig. 5.15 Schematic conceptualization of value creation (Belvedere et al. 2013)

[1]The Forrester effect (also known as Bullwhip effect) is a distribution channel phenomenon: it refers to increasing swings in inventory orders in response to shifts in customer demand, when moving upstream in the supply chain.

5.3.2 Installed Base Information

Installed base are all the solutions supplied to the customer's plant. Consider, for example, the set of clippers in a department, the whole fleet made available in logistics services. The set of data collected concerning the operation and operation of the solutions constitutes the installed base information. The set of these data represents a key resource in the development of the PSS because the knowledge allowing a correct implementation and development derives from them. For some companies, the set of solutions installed is the only asset (Wise and Baumgartner 1999; Ulaga and Reinartz 2011) making the management of such information fundamental. In other words, in the PSS world, correctly managing the installed base is fundamental since it is a source of knowledge and source of new offers of service and revenue models (Storbacka 2011). The level of control that the company exercises on the information generated by the customer during the use of the product is crucial to collect and update the historical data after each repair and maintenance and the monitoring for preventive maintenance and optimization of the processes of the customer (Neff et al. 2014) and depends on the control on the basic installed. For instance, Rolls-Royce has developed the management of systems for the installed basic information allowing the implementation of product–service logics and modifying its business model, passing from a transactional scheme to a relational one and pointing on the value in use rather than on the value in the exchange. By adopting such systems Rolls-Royce has reduced the operational risks in which it was previously incorporated. In addition to the structures linked to the collection of data in itself, the instruments used for their analysis and their interpretation are important (Saccani et al. 2014).

5.3.3 Human Resources

The servitization of the offer involves a change in the management of human resources. For example, a change of mentality and cultural restructuring are necessary in order to tune the collective sensibility on the service logic at every organizational level (Gebauer et al. 2005; Barquet et al. 2013). In this sense, companies must invest heavily in human resources in order to develop new skills and reconfigure existing ones (Ulaga and Reinartz 2011; Kindström and Kowalkowski 2014). As we have already mentioned above, the transition from a traditional context based on exclusive product sales to PSS is not a simple process. In fact, it is necessary to develop internal competences that irrevocably involve the management of our human resources (Parida et al. 2014):

- Business model design: combining products and services represents a challenge for companies that undertake the path of servitization. Above all, it is necessary to be able to structure the business model so as to understand the needs of the client and in order to transmit the value that he requires. Redefining the relationship with the customer force companies to develop a different approach in marketing, capable of giving equal weight both to the aspects related to the product and the service, and therefore requires more technical training of sales personnel. In order to correctly define the value to be transmitted, it is necessary to focus on pricing with more refined methods.

- Network management: The introduction of PSS involves a redefinition of the network and the entry of new partners. Managing the network correctly is often a matter of aligning incentives along the network (Parida et al. 2014). To ensure this alignment with various partners, companies often introduce processes to build partner knowledge and increase relational skills. With this you want to get a better understanding of the client's goals, skills and directions of growth. Some companies have started co-managed projects to identify new solutions to problems just emerging. The importance of network management has led to the creation of functional units dedicated to this aspect, responsible for managing and maintaining relations with partners.
- Integrated development: for manufacturing companies, product development still requires the company's main efforts. An integrated offer of product–service solutions requires the two elements to be managed in symbiosis already in the design phase. To correctly manage this step, it is necessary that the personnel involved in the development is aware of the operational and organizational context of the client in which the solutions will operate.
- Service delivery management: the effective operation of the solutions offered relies on a network composed of dealers, distributors, service partners and other branches that have an active role in connecting the final consumer with the upstream provider. The skills necessary to tackle this task are those of the partners involved in the installation and execution of the service components (Fig. 5.16).

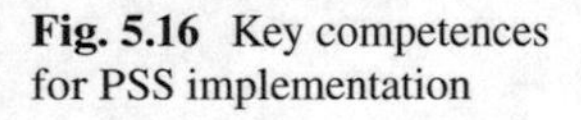

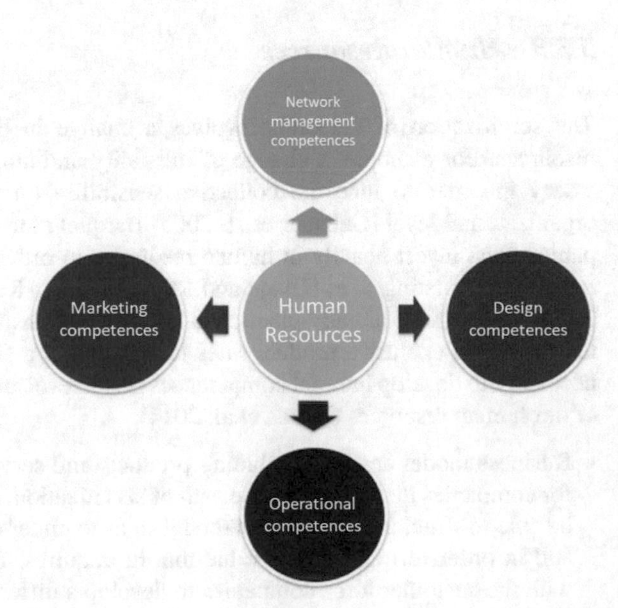

Fig. 5.16 Key competences for PSS implementation

5.3.4 *Financial Resources*

The introduction of PSS logics into a company has a profound impact on the financial resources. In fact, in traditional contexts, the manufacturing companies benefited from the frequent positive cash flows due to the purely relational nature of their business. The capital invested in PSS can be a criticality since the company can remain the owner of the product and the break-even point in time may be longer compared to the physical sale of the product (Tukker 2004; Barquet et al. 2013). In the event that a company decides to provide use-oriented solutions or result-oriented fact, the ownership of the asset may burden for many years on the budgets of providers obliging providers to provide adequate financial resources to overcome this period (Mont 2002). In addition, even expanding product and service offerings means exposing oneself to a great financial risk (Kindström and Kowalkowski 2014; Alghisi and Saccani 2015) which must be faced with adequate resources or by tightening closer relations with financial partners.

5.4 Key Partners in PSS Implementation

The supply of PSS adds new tasks to the operational activities of companies that intend to focus on this type of offering. Since it is not possible to develop the necessary set of skills to face this kind of challenges, companies must develop a suitable network and build relational structures with key partners (Baines et al. 2009; Gao et al. 2011). A network describes the relationships and interactions with the different stakeholders (customers, dealers, service partners and suppliers). As relationships become tighter, it is difficult and very expensive to maintain a high number of customers, so the process of selecting stakeholders becomes extremely important (Mont 2002). It may happen that this necessity pushes companies to collaborate also with realities belonging to unexplored contexts in order to access to new skills (Evans et al. 2007). After choosing partners and determining the level of interaction, more effort is required to develop practices for coordinating relationships and sharing the right information in the network (Schuh et al. 2009). In this segment, companies should look at the key partners actively involved in the development and implementation of PSS systems.

The type of partner varies depending on the service, but there are general aspects that intrinsic to servitization. In first analysis, it is necessary to underline how, in the process of creation and implementation of the solutions offered, the customer plays a fundamental role, becoming a key partner thanks to the mechanism of co-creation. The client projects his own experiences and needs within the design and product development process, allowing the manufacturer to add further value to his own solutions. If traditionally the customer was an external actor, in the PSS world the development of a deep relationship based on trust with the customer is a key success factor. Another key partner that takes on new nuances of role within the network of PSS world is the financial who took over the ownership of the asset or participates in joint ventures. Legislative institutions also take on a more structured role as they enact more stringent laws on the environmental impact and management of the end of life of the materials used for the components produced. In addition to these partners, it is possible to recognize others by analysing the three types of PSS: product-oriented, use-oriented and result-oriented. In the case of product-oriented services, the activities related to the delivery and installation of the asset are generally in charge to the provider who can, however, delegate them to an external dealer (Tukker 2004). This means that in some configurations, the provider does not have direct contact with the customer by virtue of the presence of a third-party partner assigned to the delivery (Reim et al. 2015). Going to the use-oriented category, it is customary to outsource the part related to the delivery both in the B2B and B2C contexts to third parties. As seen above, in the use-oriented modalities, financial institutions are involved in the customer–supplier relationship by providing the cash-flow necessary for the financial support of the transaction (Mont et al. 2006). At the end of the life cycle of the good (or of the supply contract), an additional player can be involved in disposal or recovery of equipment. The last category is the result-oriented, where the network structure changes significantly. This type of supply is close to the concept of vertical integration and the interpenetration between customer and supplier is crucial. In addition to close collaboration with the client, other stakeholders (financial institutions, transport and recycle companies) may be involved (Azarenko et al. 2009). The value proposition in the PSS embraces a vast and complex network of stakeholders. The quality of the relationship between the manufacturer and its PSS network influences the life cycle of the PSS and the client's activities. The transition to a model based on PSS requires a redefinition of the relational structure and of the actors (Mont 2002; Ferreira et al. 2013; Liu et al. 2014; Reim et al. 2015). It is essential to move from short-term price-based relationships to broader strategic relationships (Barquet et al. 2013) (Fig. 5.17).

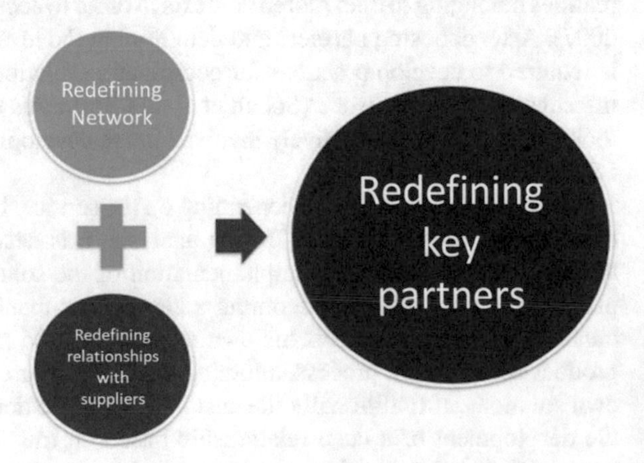

Fig. 5.17 Redefining key partners and their roles: key elements

5.4.1 Network

Implementing PSS solutions increases the operational complexity within the company (Reim et al. 2015) and companies pursuing this strategy must develop a new relational structure (Baines et al. 2009; Gao et al. 2011) in order to share capabilities and create value. The new network must therefore be redesigned taking into account the value proposition (Aurich et al. 2006), specifying the role of each partner and the value through the life-cycle (Storbacka 2011). Therefore, the creation of such a network requires the identification of the actors and of their main skills (Barquet et al. 2013). Finding partners who can add value to the new offer is a critical phase and suppliers should be selected with criteria that go beyond those based on price. The study conducted by Gebauer et al. (2013) in the context of capital goods manufacturing, illustrates four types of networks that support the implementation of PSS. They are the vertical after-sales service network, the horizontal outsourcing service network, the vertical life-cycle service network and the horizontal integration life-cycle service network. The vertical and horizontal terms describe the structure of the network (Möller et al. 2005). Although this classification is widely accepted, networks are rarely truly vertical or horizontal. The horizontal networks can, for example, also contain some vertically positioned suppliers and vice versa. The vertical–horizontal denomination therefore refers to the dominant orientation. Concepts such as after-sales services, service outsourcing, life-cycle service and integration service describe the type of service offered by the network.

Type A: Vertical After-Sales Service Network **(Fig. 5.18)**

The vertical term implies that the actors cover up or down activity in a single specific value chain. The after-sales term indicates that the activities are concentrated on the use of the product. The network includes the physical producer, logistics service providers and component suppliers. The OEM is at the centre of the network. This type of network is a fairly stable business system and covers all planned value activities. Products are highly standardized and the activities are therefore well known and predefined. The whole value proposition of the whole relational structure focuses on the post-sale phases or on the use of the product. The activities carried out by each actor enable the use of the product. The OEM offers services such as delivery of spare parts, repairs and maintenance. By storing and transporting spare parts, logistics providers support the specialist in this activity. The service network may also include external distributors or partners if the demand does not take on such proportions as to establish subsidiaries in other markets. The key partners in this configuration are specialized distributors for sale, installation and maintenance.

Type B: Horizontal Outsourcing Network **(Figs. 5.19 and 5.20)**

The horizontal notion suggests that actors cover different value chains. A specialist supported by outsourcing logics is the focal point of the network. Companies are turning towards outsourcing services for different types of equipment. There are two possible configurations for the outsourced specialist. In the first one the role is covered by the OEM, which expands its outsourcing services beyond its product categories. The specialist takes full responsibility for operating and maintaining the equipment at the customer's premises. Alternatively, the role is covered by the client, who thus expands the tasks of his maintenance service, reaching new business possibilities. The skills acquired can be coordinated as services to be outsourced and offered to other customers. Notwithstanding this small difference, the rest of the network remains substantially the same. Taking responsibility for operational and maintenance services allows outsourcing specialists to break the dyadic relationship between customer and provider. The value proposition is, therefore, subordinated to

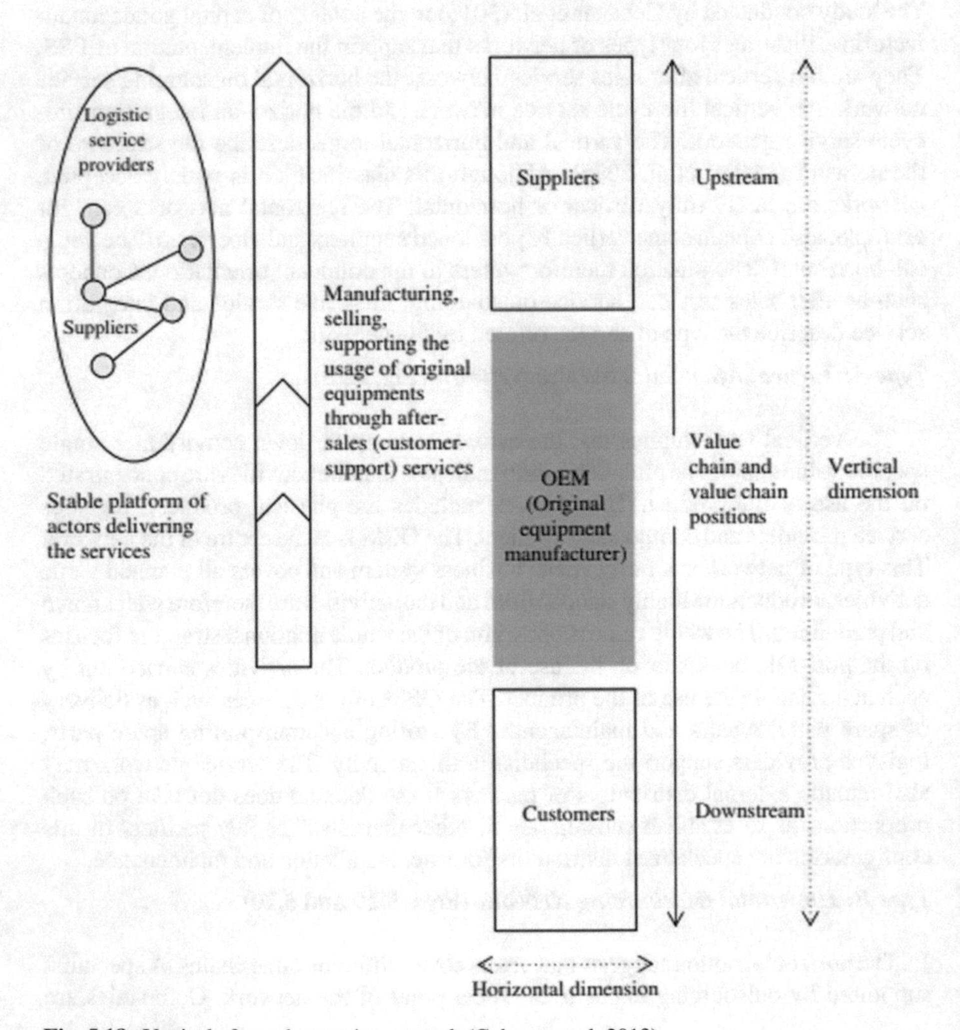

Fig. 5.18 Vertical after-sales service network (Gebauer et al. 2013)

maintenance and operational processes that are managed in a more efficient manner than individual actors can do alone. In contrast to type A, this type of network is a stable platform of actors, but it allows flexible configuration according to the client's requests. This means that the specialist selects the necessary actors and their value activities from a stable platform. This selection is based on the congruence between the client's needs and the ability of the actors and in this the outsourcing specialist plays the role of director and coordinator. In this type of network, customers prefer a mutual dependency with similar players when the activity involved is not too important a process. The other OEMs are forced to offer only basic services such as supply of spare parts, warranty services or services related to the solution of problems with a high technical thickness. Other activities such as inspections, repairs, maintenance and updating, and process optimization are delegated to external actors. Commonly with what is seen in type A, the logistics provider support the specialist in the storage and handling of spare parts. The other OEMs and upstream actors ensure the presence of such spare parts and participate in the recovery and repair activities. To ensure remote monitoring, specialists can use partners that provide IT support.

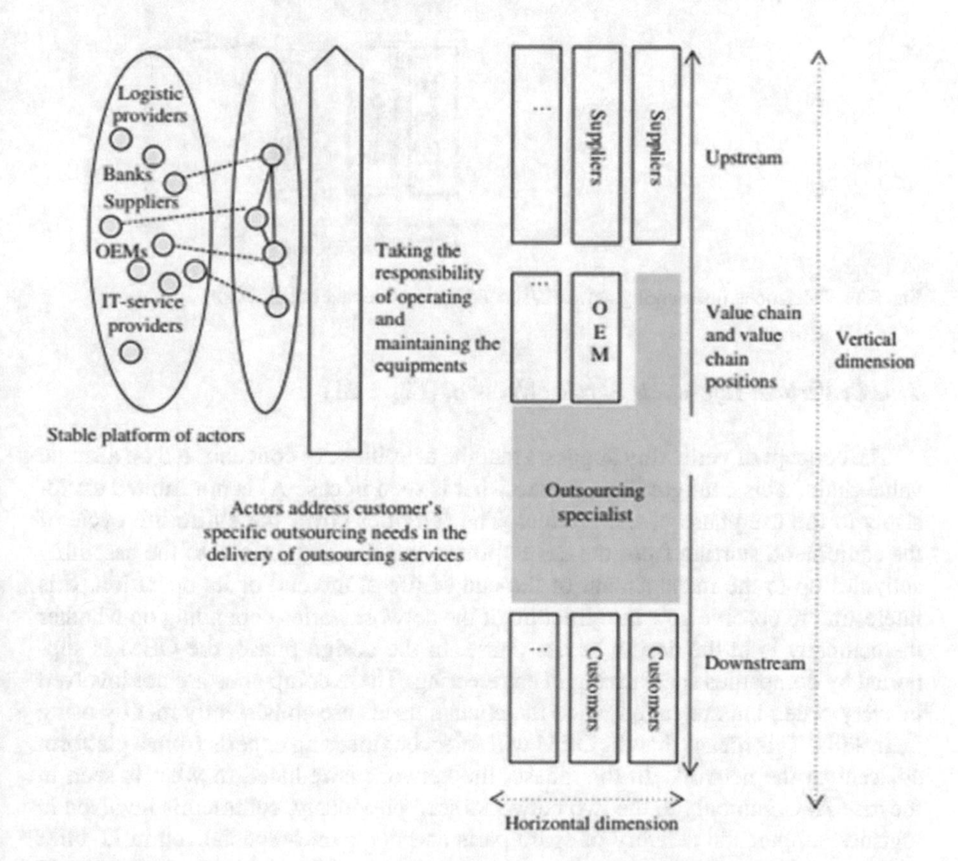

Fig. 5.19 Horizontal outsourcing network, first option (Gebauer et al. 2013)

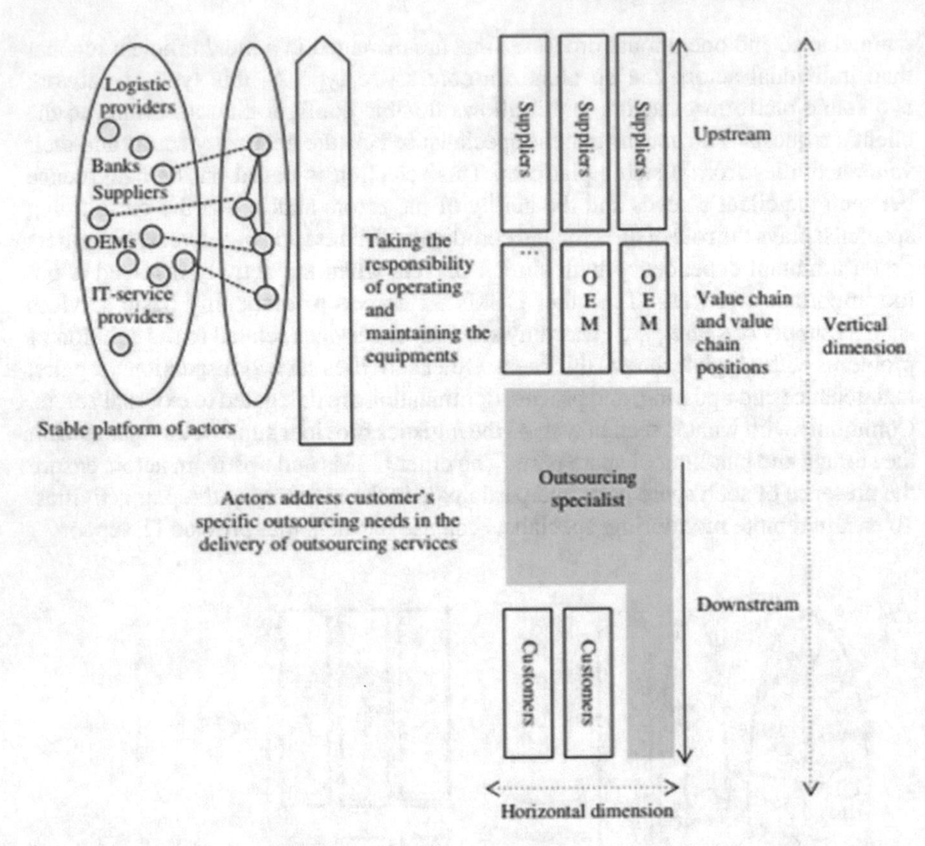

Fig. 5.20 Horizontal outsourcing network, second option (Gebauer et al. 2013)

Type C: Vertical Life-Cycle Service Network (Fig. 5.21)

The concept of verticality suggests that the activities are concentrated on a single value chain. This configuration, unlike what is seen in case A, is not limited exclusively to the use phase of the product. The activities cover the entire life cycle of the equipment, starting from the development and design to move to the assembly activities up to the management of the end of life at the end of its operation. It is interesting to observe how the structure of the network varies depending on whether the company is in the design or use phase. In the design phase, the OEM is supported by companies specializing in engineering. These companies are not involved in every order, but are called when the client's needs are consistently met by using their skills. This means that the OEM will select engineering experts from a platform adjacent to the network. In this phase, the network is reduced to what is seen in the case A. Commonly to the two networks seen previously, contractors involved in logistics support the delivery of spare parts and companies specialized in IT offer remote monitoring services.

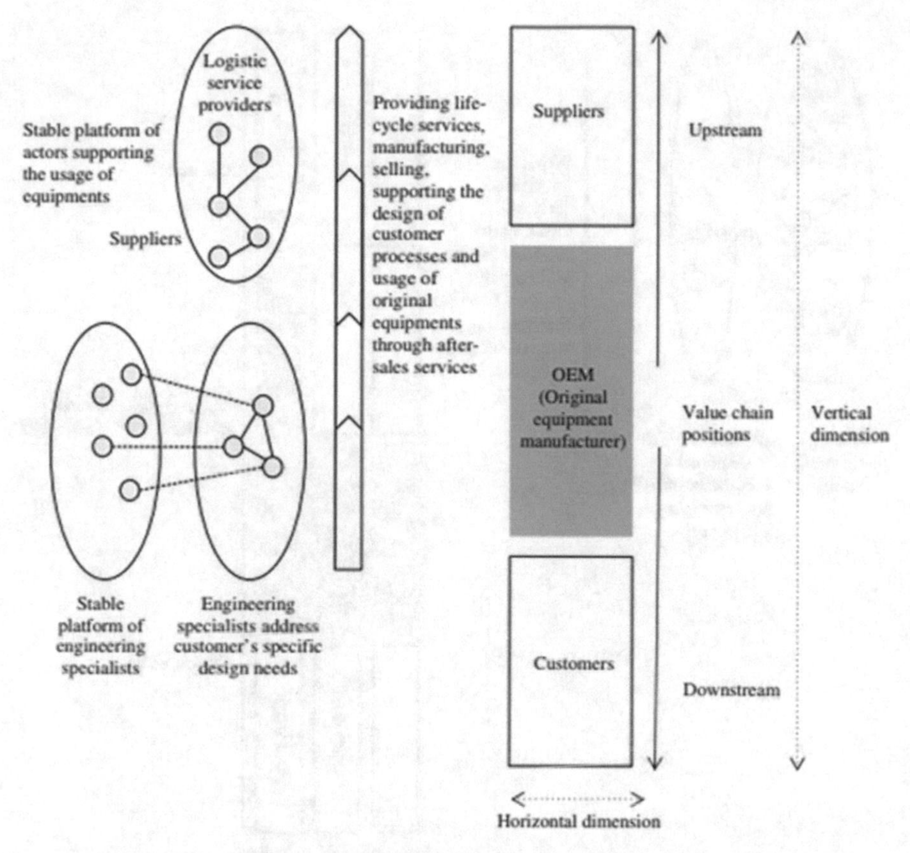

Fig. 5.21 Vertical life-cycle service network (Gebauer et al. 2013)

Type D: Horizontal Integration Service Network (Fig. 5.22)

As seen in type B, the term horizontal suggests that the assets are spread across multiple value chains. In contrast to what we saw in the first two types, but similarly to the third, horizontal integration already starts from the product development phase. The OEM represents the focal point of the network and covers the design, manufacture and maintenance of the equipment. In addition, the OEM offers services for third-party products. The final purpose of a network so configured is to integrate the various services present in a single working solution. This type of network includes a large set of auxiliary providers who contribute in building the solutions adopted. Together with suppliers in the various value chains, the auxiliary providers form a new value system. In this case, the network structure is not purely horizontal since vertical elements can be included such as strategic alliances with experienced local market partners. The OEM often forms this type of relationship with local companies that are practical in providing the requested services. The network taken in

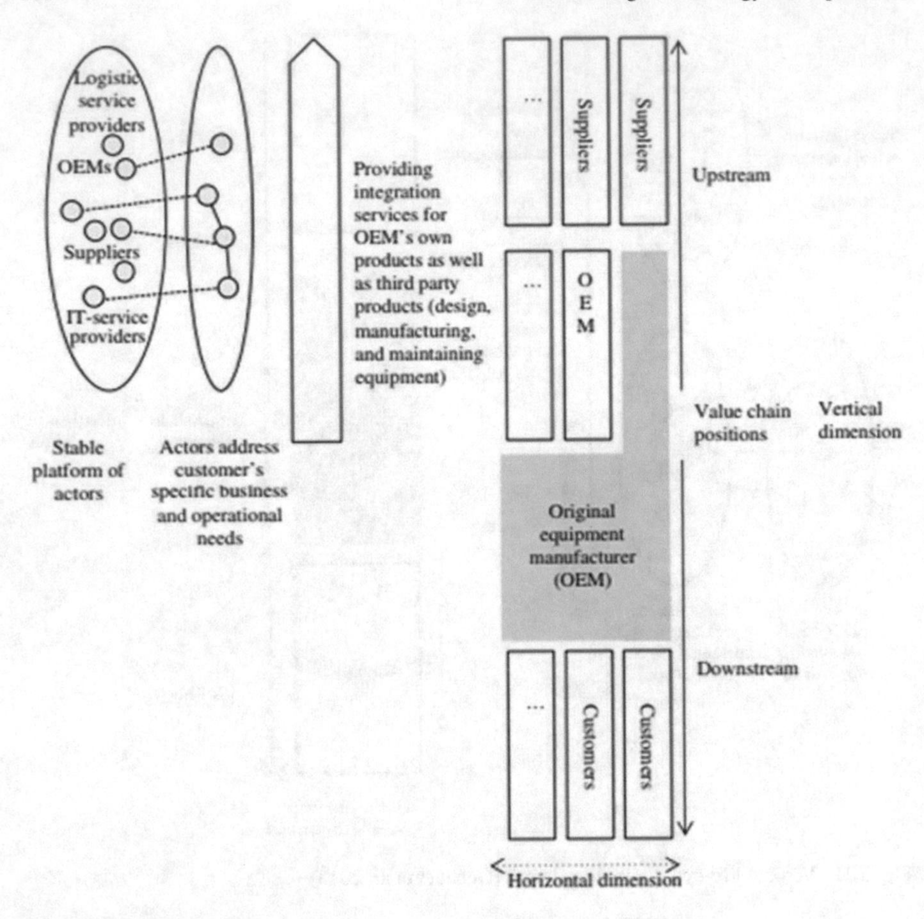

Fig. 5.22 Horizontal integration service network (Gebauer et al. 2013)

analysis forms a flexible platform in the sense that the leading company selects and orchestrates the necessary tools to solve that the network must face.

5.4.2 Supplier Relationship

In BMs based on PSS, it is necessary to develop and maintain strong relationships with critical suppliers (Gebauer et al. 2013), and actors become more and more interdependent and harmonization within and beyond the organizational boundaries of each actor involved is fundamental (Oliva and Kallenberg 2003; Brady et al. 2005). Once the partners are chosen, a major effort is required in developing practices for coordinating relationships and sharing the right information effectively in the network (Schuh et al. 2009). In essence, once the initial effort to set up long-term relationships has been profuse, this should be reduced to cooperation and meaningful sharing of bidirectional information (Saccani et al. 2014). For these reasons, particular attention should be paid to the communication interface between the partners, in order to guarantee transparency and quality in the information exchange (Storbacka 2011). In the study proposed by Saccani et al. (2014), the typology of services is used as a lens to analyse the relations between producer and customer. This approach is motivated by the differences that exist between the different types of existing PSS influencing the relationship between buyer and supplier. The three categories taken into consideration are shown below.

Product Support (PS) Services

This type includes traditional after-sales services such as installation and maintenance characterized by a lack of customization from a low cooperation between provider and user due to a lack of complexity and a transactional nature. In order to deliver these services, a deep experience of the product of the workforce involved and refined management skills are necessary. In this context, information exchange is limited to some aspects of technical and operational nature. There are several advantages related to this, however, limited information exchange, greater accountability and a greater capacity for use is an improved ability to forecast. It should not be overlooked a certain reticence by the customer to provide information that can lead to an increase in inefficiencies within the supply chain, but also to a lack of knowledge and the increase in tensions between customer and supplier. The evidence suggests that when a supplier provides only this type of services, the manufacturer exercises a resistance motivated by its bargaining power. On the other hand, the supplier limits the information provided to those established by contract. In addition, operational links smooth processes and increase efficiency, in particular through integrated information systems, which allow parties to share schedules, request spare parts and provide feedback on field operations. The research shows that the operational links are imposed on the buyer and may not benefit the supplier. In order to avoid opportunistic behaviour, buyers adopt legal measures that restrict the relationship to standards. The more the service is standardized the more this type of control is effective.

Customer Support (CS) Services

The degree of complexity and customization of this type of services can vary significantly depending on the business context. However, their hallmark is their significant level of interaction with the final user. Since this type of service is based on the interaction with the client, to be implemented in the correct way it is necessary to focus on the latter. This goes beyond the traditional skills of the manufacturer leading to strategies related to outsourcing. The exchange of technical information is similar to the services seen previously, but is paid by the customer who returns feedback on quality and further suggestions. The operational links consist of systems that automate the information flow between the parts and interface the personnel of both sides. Formal meetings are organized in the areas where values and objectives are shared. The customer checks the quality of the performance comparing them to what is defined in the contract phase and claims the penalties if these do not conform to the standards. This type of relationship is usually prolonged over time and leads to a real cooperation aimed at satisfying market needs.

Process-Related (PR) Services
This kind of services is usually very complex and highly personalized. These services are designed to respond to specific needs and to improve processes related to the product offered. The supply of this type of package obliges the supplier to acquire a deep knowledge about the product, the individual needs of the single users and the economic and organizational context in which the goods are inserted. The information exchanged is many and bidirectional, providing great knowledge regarding the customer's perception of the value. Both parties can enjoy the benefits of this exchange of knowledge which has recently led to new payment mechanisms on the pay-per-use model. With the thickening of the relationship, a deep normative formulation of this cooperation takes place. Operational links are totally automated and the meetings organized are of a strategic nature and focus on long-term aspects. Legal constraints become softer and a more trust-based relationship emerges.

Seven Key Facts

- The adoption of product service system impacts on the whole company's business model and on operations strategy and operations management as well.
- Various methods can be adopted for PSS design, combining elements of product and service design.
- Designing product and service configuration and delivery are other two key activities in ensuring a successful design of the entire PSS-related business model.
- PSS design also requires formal processes and methods to realize intra-functional activities and inter-functional integration.
- A well-designed mix of physical resources, financial resources and human resources is at the core of an effective operations strategy for PSS.
- Human resources must cover a wide range of competences: network management, design, marketing and operational competences.
- PSS implementation requires a redefinition of key partners, for what concerns relationships and network structure.

References

A. Alghisi, N. Saccani, Internal and external alignment in the servitization journey—overcoming the challenges. Prod. Planning Control **26**(14–15), 1219–1232 (2015)

T. Alonso-Rasgado, G. Thompson, B.O. Elfström, The design of functional (total care) products. J. Eng. Des. **15**(6), 515–540 (2004)

M. Ardolino, M. Rapaccini, N. Saccani, P. Gaiardelli, G. Crespi, C. Ruggeri, The role of digital technologies for the service transformation of industrial companies. Int. J. Prod. Res. **56**(6), 2116–2132 (2018)

J.C. Aurich, C. Fuchs, C. Wagenknecht, Life cycle oriented design of technical product-service systems. J. Clean. Prod. **14**, 1480–1494 (2006)

J.C. Aurich, N. Wolf, M. Siener, E. Schweitzer, Configuration of product-service systems. J. Manuf. Technol. Manage. **20**(5), 591–605 (2009)

A. Azarenko, R. Roy, E. Shehab, A. Tiwari, Technical product-service systems: Some implications for the machine tool industry. J. Manufact. Technol. Manag. **20**(5), 700–722 (2009)

T.S. Baines, H.W. Lightfoot, O. Benedettini, J.M. Kay, The servitization of manufacturing: a review of literature and reflection on future challenges. J. Manufact. Technol. Manage. **20**(5), 547–567 (2009)

A.P.B. Barquet, M.G. de Oliveira, C.R. Amigo, V.P. Cunha, H. Rozenfeld, Employing the business model concept to support the adoption of product-service systems (PSS). Ind. Mark. Manage. **42**(5), 693–704 (2013)

J. Becker, D. Beverungen, R. Knackstedt, M. Matzner, O. Muller, J. Poppelbuss, Bridging the gap between manufacturing and service through it-based boundary objects. IEEE Trans. Eng. Manage. **60**(3), 468–482 (2013)

V. Belvedere, A. Grando, P. Bielli, A quantitative investigation of the role of information and communication technologies in the implementation of a product-service system. Int. J. Prod. Res. **51**, 410–426 (2013)

S. Bonetti, M. Perona, N. Saccani, Total cost of ownership for product-service system: application of a prototypal model to aluminum melting furnaces. Procedia CIRP **47**, 60–65 (2016)

T. Brady, A. Davies, D. Gann, Creating value by delivering integrated solutions. Int. J. Project Manage. **23**, 360–365 (2005)

F. Buttle, L. Ang, R. Iriana, Sales force automation: review, critique, research agenda. Int. J. Manage. Rev. **8**(4), 213–231 (2006)

S. Cavalieri, G. Pezzotta, Product-service systems engineering: state of the art and research challenges. Comput. Ind. **63**, 278–288 (2012)

M. Cook, T.A. Bhamra, M. Lemon, The transfer and application of product service systems: from academia to UK manufacturing firms. J. Clean. Prod. **14**, 1455–1465 (2006)

A. Dimache, T. Roche, A decision methodology to support servitization of manufacturing. Int. J. Oper. Prod. Manage. **33**(11–12), 1435–1457 (2013)

S. Evans, P.J. Partidário, J. Lamberts, Industrialization as a key element of sustainable product-service solutions. Int. J. Prod. Res. **45**(18–19), 4225–4246 (2007)

F.N.H. Ferreira, J.F. Proenca, R. Spencer, B. Cova, The transition from products to solutions: external business model fit and dynamics. Ind. Mark. Manage. **42**(7), 1093–1101 (2013)

W.W. Fisher, *Promises to Keep: Technology, Law, and the Future of Entertainment* (Stanford Law and Politics, Stanford, CA, 2004)

J. Gao, Y. Yao, V.C.Y. Zhu, L. Sun, L. Lin, Service-oriented manufacturing: a new product pattern and manufacturing paradigm. J. Intell. Manuf. **22**(3), 435–446 (2011)

H. Gebauer, E. Fleisch, T. Friedli, Overcoming the service paradox in manufacturing companies. Eur. Manage. J. **23**(1), 14–26 (2005)

H. Gebauer, M. Paiola, N. Saccani, Characterizing service networks for moving from products to solutions. Ind. Mark. Manage. **42**(1), 31–46 (2013)

C. Grönroos, A service perspective on business relationships: the value creation, interaction and marketing interface. Ind. Mark. Manage. **40**(2), 240–247 (2011)

M. Holweg, The three dimensions of responsiveness. Int. J. Oper. Prod. Manage. **25**(7), 603–622 (2005)

D. Kindström, C. Kowalkowski, Development of industrial service offerings: a process. J. Serv. Manage. **20**(2), 156–172 (2009)

D. Kindström, C. Kowalkowski, Service innovation in productcentric firms: a multidimensional business model perspective. J. Bus. Ind. Mark. **29**(2), 96–111 (2014)

D. Kindström, C. Kowalkowski, T.B. Alejandro, Adding services to product-based portfolios: an exploration of the implications for the sales function. J. Serv. Manage. **26**(3), 372–393 (2015)
H. Komoto, T. Tomiyama, Integration of a service CAD and a life cycle simulator. CIRP Ann-Manufact. Technol. **57**(1), 9–12 (2008)
G. Lay, M. Schroeter, S. Biege, Service-based business concepts: a typology for business-to-business markets. Eur. Manage. J. **27**(6), 442–455 (2009)
C.H. Liu, M.-C. Chen, Y.-H. Tu, C.-C. Wang, Constructing a sustainable service business model: an S-D logic-based integrated product service system. Int. J. Phys. Distrib. Logistics Manage. **44**(1–2), 80–97 (2014)
N. Maussang, P. Zwolinski, D. Brissaud, Product-service system design methodology: from the PSS architecture design to the products specifications. J. Eng. Des. **20**(4), 349–366 (2009)
H. Meier, R. Roy, G. Seliger, Industrial product-service system—IPS2. CIRP Ann-Manufact. Technol. **59**, 607–627 (2010)
K. Möller, A. Rajala, S. Svahn, Strategic business nets—Their type and management. J. Bus. Res. **58**(9), 1274–1284 (2005)
O. Mont, Clarifying the concept of product-service system. J. Cleaner Prod. **10**, 237–245 (2002)
O. Mont, C. Dalhammar, N. Jacobsson, A new business model for baby prams based on leasing and product remanufacturing. J. Clean. Prod. **14**(17), 1509–1518 (2006)
N. Morelli, Designing product/service Systems: a methodological exploration. Des. Issues **18**(3), 3–17 (2002)
A.A. Neff, F. Hamel, T.P. Herz, F. Uebernickel, W. Brenner, J. vom Brocke, Developing a maturity model for service systems in heavy equipment manufacturing enterprises. Inf. Manage. **51**(7), 895–911 (2014)
F. Nordin, C. Kowalkowski, Solutions offerings: a critical review and reconceptualization. J. Serv. Manage. **21**(4), 441–459 (2010)
R. Oliva, R. Kallenberg, Managing the transition from products to services. Int. J. Serv. Ind. Manage. **14**(2), 160–172 (2003)
V. Parida, D. Sjodin Ronnberg, J. Wincent, M. Kohtamaki, Mastering the transition to product-service provision: insights into business models, learning activities, and capabilities. Res. Technol. Manage. **57**(3), 44–52 (2014)
K.S. Pawar, A. Beltagui, J.C. Riedel, The PSO triangle: designing product, service and organisation to create value. Int. J. Oper. Prod. Manage. **29**(5), 468–493 (2009)
M. Rapaccini, N. Saccani, G. Pezzotta, T. Burger, W. Ganz, Service development in product-service systems: a maturity model. Serv. Ind. J. **33**(3–4), 300–319 (2013)
M. Rapaccini, F. Visintin, Devising hybrid solutions: an exploratory framework. Prod. Plann. Control **26**(8), 654–672 (2015)
W. Reim, V. Parida, D. Örtqvist, Product-service systems (PSS) business models and tactics–a systematic literature review. J. Clean. Prod. **97**, 61–75 (2015)
N. Saccani, F. Visintin, M. Rapaccini, Investigating the linkages between service types and supplier relationships in servitized environments. Int. J. Prod. Econ. **149**, 226–238 (2014)
T. Sakao, Y. Shimomura, E. Sundin, M. Comstock, Modeling design objects in CAD system for service/product engineering. Comput. Aided Des. **41**, 197–213 (2009)
G. Schuh, W. Boos, S. Kozielski, Lifecycle Cost-orientated Service Models for Tool and Die Companies. In *Proceedings of the 1st CIRP Industrial Product-Service Systems (IPS2) Conference*, 249 (Cranfield University Press 2009)
D. Simmons, *Field Service Management: A Classification Scheme and Study of Server Flexibility*. Ph.D. Thesis. Binghamton University, Binghamton, NY, 2001
K. Storbacka, A solution business model: capabilities and management practices for integrated solutions. Ind. Mark. Manage. **40**(5), 699–711 (2011)
K. Storbacka, C. Windahl, S. Nenonen, A. Salonen, Solution business models: transformation along four continua. Ind. Mark. Manage. **42**(5), 705–716 (2013)
E. Sundin, B. Bras, Making functional sales environmentally and economically beneficial through product remanufacturing. J. Clean. Prod. **13**, 913–925 (2005)

A.R. Tan, T.C. Mcaloone, Characteristics of strategies in product– service-system development. *Proceedings of the 8th International Design Conference*, 2006, pp. 1–8

A. Tan, T.C. McAloone, L.E. Hagelskjær, Reflections on product/service-system (PSS) conceptualization in a course setting. Int. J. Des. Eng. (2009)

T. Tomiyama, Service engineering to intensify service contents in product lifecycles. In *Proceedings of Eco-design*, 2001, pp. 613–618

A. Tukker, Eight types of product-service system: eight ways to sustainability? Experience from SusProNet. Bus. Strategy Environ. **13**, 246–260 (2004)

W. Ulaga, W.J. Reinartz, Hybrid offerings: how manufacturing firms combine goods and services successfully. J. Mark. **75**(6), 5–23 (2011)

G.V.A. Vasantha, R. Roy, A. Lelah, D. Brissaud, A review of product-service systems design methodologies. J. Eng. Des. **23**(9), 635–659 (2011)

E.G. Welp, H. Meier, T. Sadek, K. Sadek, Modelling approach for the integrated development of industrial product-service systems. *The 41st CIRP Conference on Manufacturing Systems*, 2008

R. Wise, P. Baumgartner, Go downstream: the new profit imperative in manufacturing. Harvard Bus. Rev. **77**(5), 133–141 (1999)

Chapter 6
How Product Service System Can Disrupt Companies' Business Model

The chapter starts presenting the PSS as a business model and adopting the theoretical frameworks of Business Model Canvas (Osterwalder and Pigneur 2010) and Business Model Innovation Process (Adrodegari et al. 2018) . Then, we introduce the six key elements of a PSS Business Model. Illustrative cases are presented to exemplify PSS implementation in different contexts, to highlight different key elements and areas impacted.

6.1 Decomposing PSS Business Models

For companies, the development of PSS can entail the opening of new market horizons, maintain competitiveness, provide an impetus to innovation and guarantee tax benefits (Mont 2002). If, for some companies, the transition is interpreted with a simple addition of services to existing products, for others this step assumes strategic connotations. The implementation of PSS allows to oust some competitors due to the difficulty in imitating a highly specific and customized package (Annarelli et al. 2016). The nature of the PSS also makes it possible to increase the differentiation of products offered (Baines et al. 2007) and, therefore, to approach in a more proactive way the market. The exclusion of the competitors is guaranteed by the constraints to which the customer is subjected, such important and difficult constraints to weave that encourage a focus on the most profitable customers (Wise and Baumgartner 1999).

As regards consumers, the changes introduced by the product service system philosophy can help to develop policies aimed at supporting sustainable lifestyles and consumption. In addition to environmental sustainability, the spread of servitization is expected to lead to the creation of new jobs, due to the fact that the permanent need for maintenance and updating of the systems for reuse and disposal may require, in the early periods, more workforce compared to the old production–sales system (Mont 2002).

A. Annarelli et al., *The Road to Servitization*,
https://doi.org/10.1007/978-3-030-12251-5_6

Consumers derive various advantages from the development of PSSs since they can add new solutions to existing ones. They have new ways to access products and services that are outside the property but, above all, they can enjoy greater consideration by the provider in terms of design and after-sales assistance. It is intuitive to imagine, referring to our daily experience, how much an adequate assistance in the use of technological products can affect customer satisfaction allowing to exploit every function of the product purchased. Likewise, it can be observed how the PSS are conceived, designed and managed in such a way as to prolong life cycle to the maximum, and the continuous maintenance and updating of the systems allow to maintain product performance (Cook et al. 2006; Armstrong et al. 2015)

Although the concept of a business model dates back to the 1950s, research on this topic has accelerated only in recent years. Business model is the logical basis for how an organization creates, transmits and receives value (Osterwalder and Pigneur 2010). More in detail the business model is the conceptual translation of an organization of three key aspects (Osterwalder 2004):

- How the key components, functions and parts are integrated in order to convey value to the customer;
- How these parts are interconnected within and, through the supply chain, with the network of stakeholders;
- How the company creates value or profit by exploiting these interconnections.

A business model should pursue the alignment of high-level strategies with the underlying assets and, thus, guarantee a competitive advantage. If the set of relationships is tacitly understood within the organization (Teece 2010), the business model will become a tool to make these interconnections explicit (Chesbrough 2010; Amit and Zott 2010) opening up new possible interpretations on how the company creates value.

6.1.1 Business Model Canvas

A well recognized model, able to describe the organizations and support the implementation of the PSS, is the Business Model Canvas proposed by Osterwalder and Pigneur (2010). This model consists of nine key elements as follows:

- Customer segment: group of people or organizations that a company wants to reach and serve;
- Value proposition: products and services that create value for a specific market segment and customers;
- Distribution channels: company interfaces with customers;
- Customer relationship; types of relationships that a company establishes and maintains with a specific customer segment;
- Revenue streams: revenue that the company obtains from each individual customer segment;

- Key resources: goods needed to offer and deliver the previous items;
- Key activities: activities involved in the development and supply of the aforementioned elements;
- Key partner: network of suppliers and partners that supports the execution of the business;
- Cost structure.

The feasibility of PSS also depends on the introduction of new techniques and new conceptualizations. The main purpose of this section is to comprehend the transition to the Product–Service logic and to highlight a series of key questions for the management that can be used as guidelines for implementing the PSSs within a business model. First there is the need to define the context in which the company will operate. The PSS must be adapted to the previous reality, so as not to create a too heavy impact in the organization and to compare the next solution to the previous one in terms of performance and customer satisfaction (Barquet et al. 2013). The process should also be gradual and risks and barriers should be calculated in advance.

Furthermore, distinctive elements of PSS (see previous chapters) can be summarized and inserted in the framework presented above, as in Figs. 6.1 and 6.2.

The BM Canvas can also be adopted as a tool for analysing and comparing different PSS alternatives. Barquet et al. (2013) provided an example for this usage of the framework to compare four different alternatives of PSS implementation in the thermoforming machines business. The Canvas can be used to represent either an

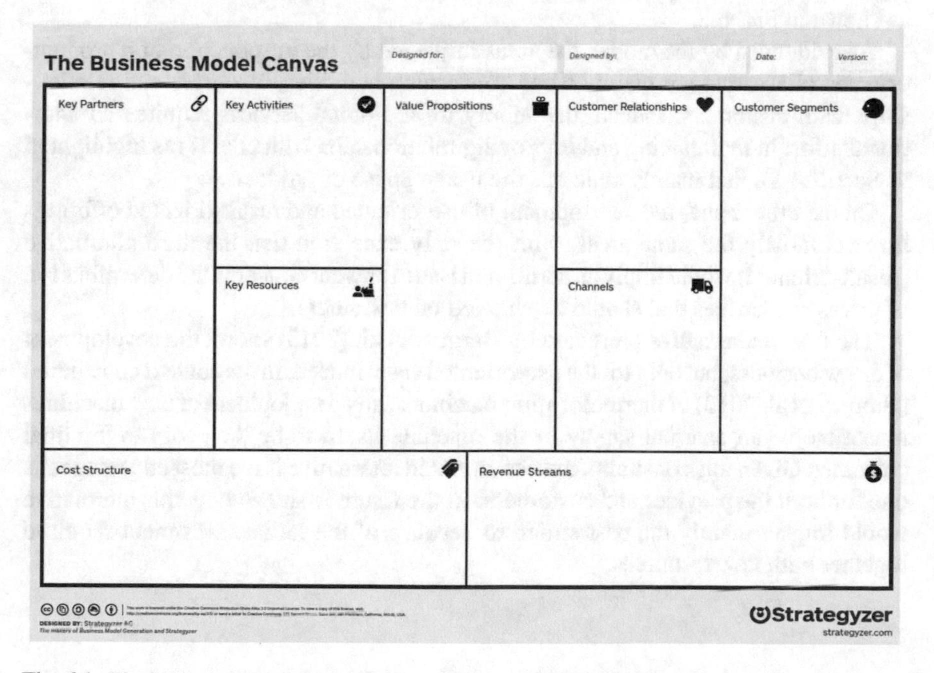

Fig. 6.1 The business model canvas (Osterwalder and Pigneur 2010)

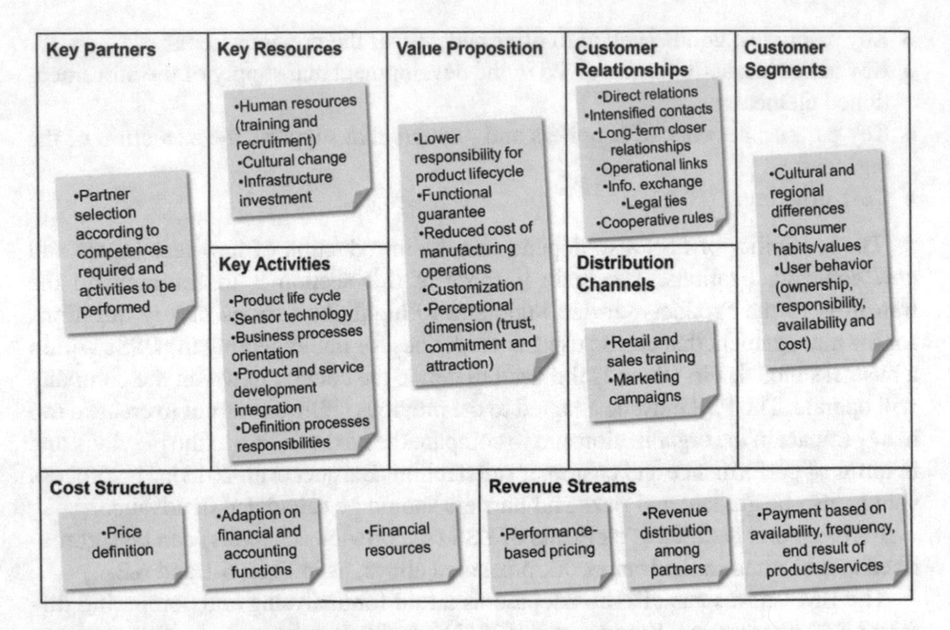

Key Partners
•Partner selection according to competences required and activities to be performed
Key Resources
•Human resources (training and recruitment)
•Cultural change
•Infrastructure investment
Key Activities
•Product life cycle
•Sensor technology
•Business processes orientation
•Product and service development integration
•Definition processes responsibilities
Value Proposition
•Lower responsibility for product lifecycle
•Functional guarantee
•Reduced cost of manufacturing operations
•Customization
•Perceptional dimension (trust, commitment and attraction)
Customer Relationships
•Direct relations
•Intensified contacts
•Long-term closer relationships
•Operational links
•Info. exchange
•Legal ties
•Cooperative rules
Distribution Channels
•Retail and sales training
•Marketing campaigns
Customer Segments
•Cultural and regional differences
•Consumer habits/values
•User behavior (ownership, responsibility, availability and cost)
Cost Structure
•Price definition
•Adaption on financial and accounting functions
•Financial resources
Revenue Streams
•Performance-based pricing
•Revenue distribution among partners
•Payment based on availability, frequency, end result of products/services

Fig. 6.2 PSS elements divided in the BM Canvas blocks (Barquet et al. 2013)

upgrade for an existing offering or the development of a new business concept/idea, as shown in Fig. 6.3.

As evidenced by the model, the areas impacted by the introduction of a product-oriented alternative are mainly those concerning activities, resources and relationships with customers. Indeed, the supply of additional services requires an additional effort in maintaining and improving relationships with clients (as highlighted in Sect. 2.4.1), that mainly concerns the usage phase of products.

On the other hand, the development of use-oriented and result-oriented offerings impacts mostly the same areas, with the only exception that the third alternative (result-oriented) would imply an additional issue for what concerns the determination of prices and/or fees that should be charged on customers.

The fourth alternative presented by Barquet et al. (2013) shows the development of a new business, but only for the use-oriented case: indeed, in the context considered (Barquet et al. 2013) of thermoforming machines, any development of new machines must involve an external supply of the machine itself, to be then sold to the final customer. Given this constraint, the use-oriented alternative is the most advantageous one for both the provider and customers. In the figure is shown how this alternative would impact mainly the cost structure because of the initial investment required together with new resources.

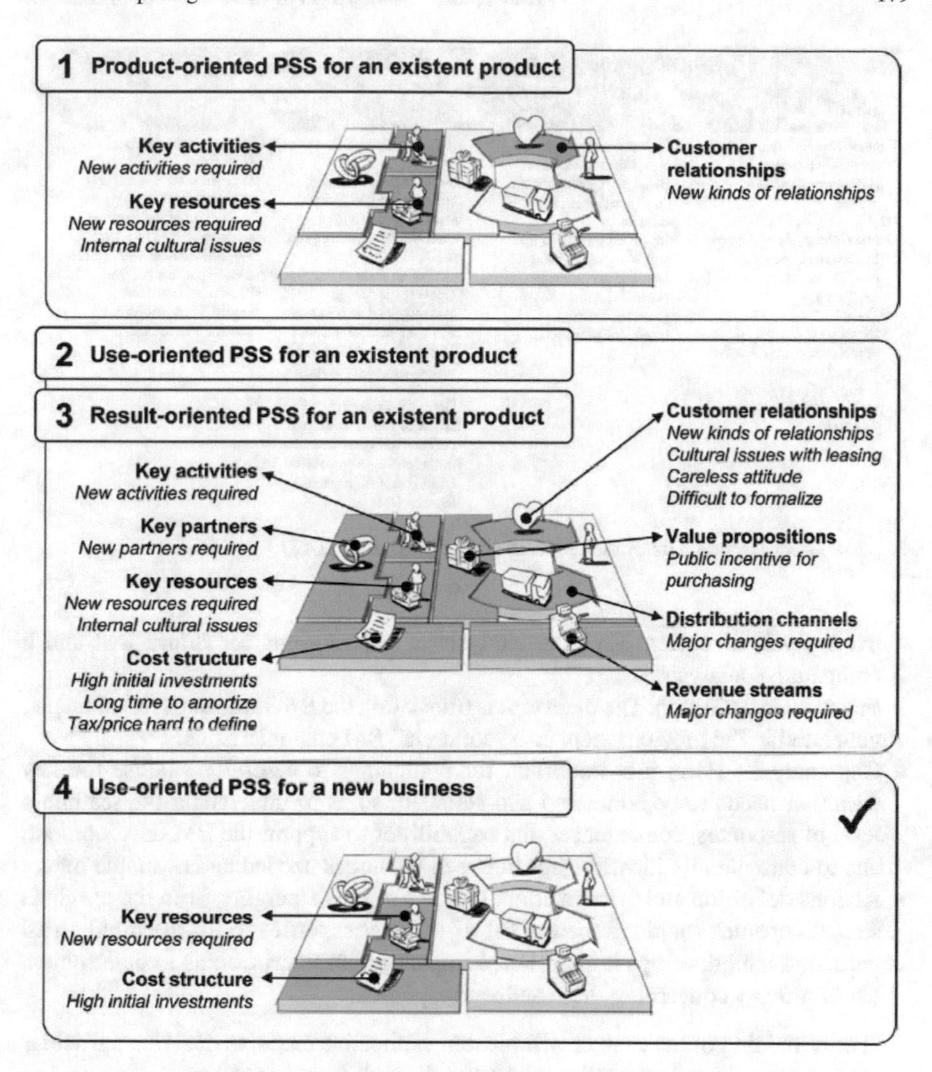

Fig. 6.3 Comparison of different PSS alternatives with the BM Canvas (Barquet et al. 2013)

6.1.2 *Business Model Innovation Process*

As evidenced throughout all the book, designing and implementing a PSS as a whole business model is a complex and challenging process. The Business Model Innovation Process framework (Fig. 6.4) can be applied to support this process, as suggested by Adrodegari et al. (2018). It is divided into four main steps, which are as follows:

- BM Idea generation: In this step, the main aim is that of defining scope and objectives behind the BM to be adopted, through three different steps that concern

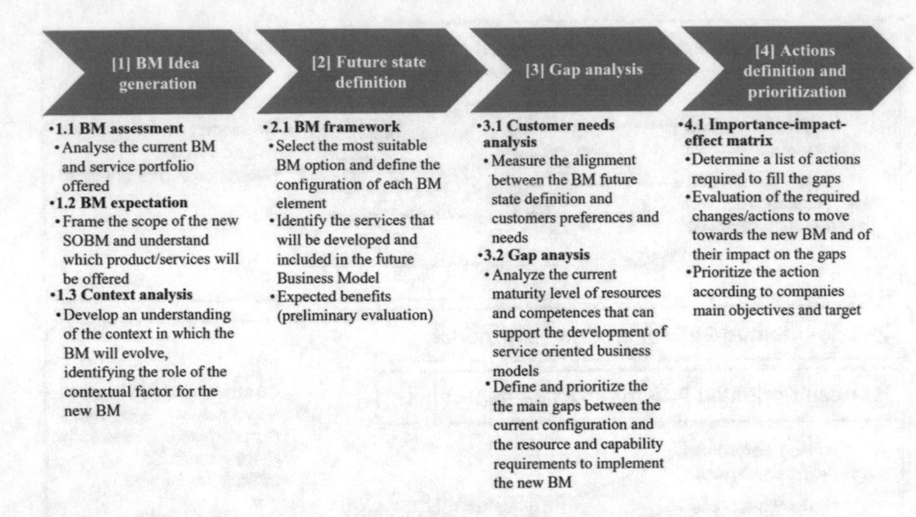

Fig. 6.4 Business model innovation process (Adrodegari et al. 2018)

the analysis of current BM, the analysis of expectations for future BM and a comparative analysis.

- Future state definition: The main task is translating the BM ideas and BM concepts, generated in the previous step, into “concrete” BM characteristics.
- Gap analysis: Here, it is important for companies to carefully analyse the key customer needs to be addressed and satisfied, so as to understand the readiness level of resources, competences and capabilities to support the BM development, and to consequently identify gaps between resources needed and available ones.
- Actions definition and prioritization: Given the gaps emerging from the previous step, companies should prepare a list of actions/responses to fill the highlighted gaps, and then develop an importance-impact-effect matrix, so as to determine a list of priority concerning these actions/responses.

The remaining of the chapter will present explicative cases, to describe particular development and implementation cases of PSS in different contexts.

6.2 Key Elements of Product Service System Business Models

PSS brings a considerable amount of novelty in affecting and transforming “traditional” business models introduces a variety of elements that can determine a considerable shift in offerings.

Figure 6.5 shows how value shifts have impacted on trends for what concerns products, manufacturing and quality. The focus is now going towards the concepts

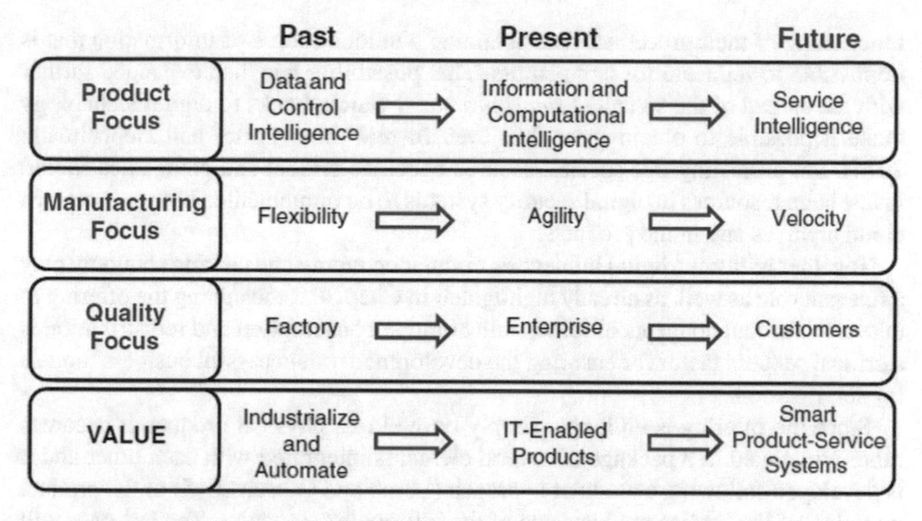

Past
Present
Future
Product Focus
Data and Control Intelligence
Information and Computational Intelligence
Service Intelligence
Manufacturing Focus
Flexibility
Agility
Velocity
Quality Focus
Factory
Enterprise
Customers
VALUE
Industrialize and Automate
IT-Enabled Products
Smart Product-Service Systems

Fig. 6.5 The evolution of value and changes in focus (Lee and AbuAli 2011)

of "velocity" and "smart" production, with quality built on a customer-centric logic (Lee and AbuAli 2011).

We can therefore state that the focus is no longer on the product itself, but it is rather going beyond the product towards newer concepts of service intelligence.

Clarifying how this shift is happening, and how it is affecting and shaping the PSS-related context, is the first step in the road to servitization.

The *six key elements* that characterize a PSS-based business model are as follows:

- Value creation
  - Design of the offering,
  - Value co-creation,
  - Functional integration with partners,
- Value delivery
  - Degree of servitization,
  - Pre- and post-sale value communication,
- Value capture
  - Short-term and long-term commitments and retention of customers.

The six key elements can be distinguished according to their contribution to value creation, delivery or capture, which are the main aims behind a business model.

Design of the offering

Thanks to the presence of services within their offer, companies can now keep a window open on their customers collecting data on the use of their assets and on the functioning of their processes, thus obtaining a unique source of information that is impossible to replicate for competitors. This possibility has then expanded further with the spread of the so-called *smart products*, which thanks to digital technology make it possible to obtain these data even in real time (Porter and Heppelmann 2014). So, protecting this special resource becomes critical and companies should invest huge resources in digital security systems to be implemented both in their own cloud archives and in the products.

Together with new digital influences, circular economy and sharing economy play a relevant role as well, as already highlighted in Chap. 4. Redesigning the offering to take into account concepts of reuse, collaborative consumption and redistribution is a critical success factor for ensuring the development of successful business models for servitization.

Since the offering is no longer simply limited to a physical product, it becomes more articulated in a package of several elements integrated with each other and it is developed following a modular approach (Lerch and Gotsch 2015) of the product modules, of the service modules and of the information modules. The last ones will have the role of connecting the first two, monitoring the performances and, of course uploading these data on the network archives of the provider. The advantages that come from this modularity are multiple: first of all, they guarantee great flexibility allowing the provider to develop a personalized offer both in scale and in scope (Cenamor et al. 2017). Furthermore, modular design reduces development costs, as a standardized module can be used for multiple customers with minimal expense while, at the same time, reducing the complexity of managing the PSS offer, which represents one of the obstacles to the implementation of a *service-centric approach*.

Value Co-creation

One of the major disruptive elements between an organization that adopts a *product-centric* proposal and one that instead proposes a product service system is the evolution in the process of value creation; this, though representing an abstract concept and not directly linked to a specific company process, critically impacts on the whole set of company activities and it is the basis of the competitive advantage guaranteed by the offer of an integrated package of products and services.

Unlike the product-focused companies, where the provider generates the value of its offer alone through the production process of the product and limits services to a role of mere support, in the case of adoption of PSS the services generate most of the value of the package proposed to the customer. The value is a function of the interactions that are generated between the two parts, such as the integration of resources and application of skills; we speak in this context, not surprisingly, of *co-creation of value* (Grönroos and Voima 2013) (Fig. 6.6).

This innovative process of creating value has been recently theorized. At the moment three enabling mechanisms have been defined:

- *Perceptual mechanism*: the ability of companies to identify, analyse and satisfy the specific needs of customers (Lenka et al. 2017). This mechanism often involves digital capabilities in implementation, as, for example, thanks to the use of sensors within the networked machinery, the provider can now have the opportunity to

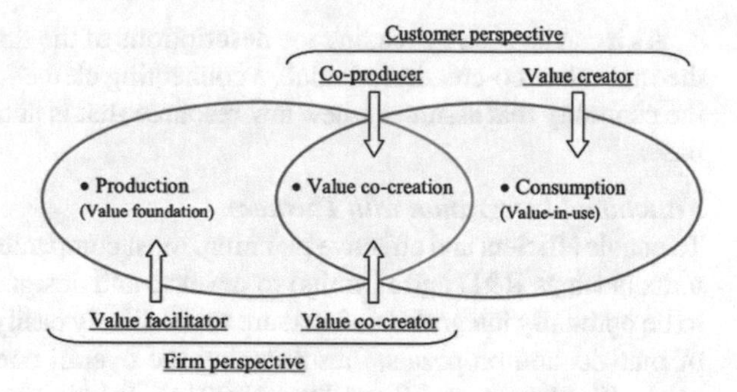

Fig. 6.6 Value co-creation mechanism enabled by PSS (Grönroos and Voima 2013)

observe the activities of his client and sharing the information obtained can help it to increase efficiency and effectiveness in the use of his product. Moreover, from the data obtained, the company can extrapolate additional indications on how to further customize the offer by structuring the PSS in the most congenial way to the customer's needs.

- *Responsive mechanism*: the ability of companies to react effectively and efficiently to evolving customer needs (Lenka et al. 2017). Also, in this case, digital capabilities play an important role in enabling or enhancing the possibilities of companies: in the current context, characterized by complex and constantly changing markets, customers are looking for providers capable of satisfying their needs in a timely and agile manner providing them with complementary features in a short time and helping them to develop proactive strategies to capitalize on any emerging value creation opportunities. This is now made possible through digital data analysis tools and shared cloud technology platforms. Another factor that guarantees elasticity is the possibility offered by PSS to develop contracts with flexible revenue models, in which the two actors often share risks and profits, also making the costs of the relationship scalable depending on the functionality required by the customer.
- Value communication mechanism: as highlighted by several authors (Baines and Lightfoot 2013; Kindström and Kowalkowski 2009), the ability to transmit the value provided to the customer is crucial in this context, given the intangibility of services, advantages and the results not always clearly perceived. In this regard, a statement by one of the interviewees present in Baines and Lightfoot's work can be emblematic: "if the client does not see what he is getting, he thinks he is not receiving anything". This capacity must, therefore, be carefully cultivated, and often takes the form of the development of *custom measurement systems* for each client, which allows them to fully appreciate the service received. Responsible for this process are those responsible for the post-sale phase of the provider, who establish a privileged relationship with the customer and exploit advanced data analysis tools to extract useful information (Kindström and Kowalkowski 2009).

As it can be seen by reading the descriptions of the three mechanisms that realize the innovative co-creation of value, a connecting element, present in each of them, is the centrality that assumes a new key resource that is data coming from the installed base.

Functional Integration with Partners

To enable efficient and effective planning, most companies proceed to the back-office units (such as R&D and IT units) to develop and design standardized modules able to be optimally integrated and that are immediately easily adaptable to various types of markets and proposals, thus reducing the overall commitment and costs of this phase (Kindström and Kowalkowski 2014). In this process, the main value chain actors are often involved, such as distributors, external service providers and core suppliers. A crucial input for this process is once again the information coming from the front-office units located further down the value chain and in direct contact with the customer and the base already installed: the two types of units must therefore cooperate closely contact to maximize the value of future innovations and, often, a digital platform is developed for fostering data circulation and knowledge sharing of best practices for managing common processes and activities (Cenamor et al. 2017) (Fig. 6.7).

Once the individual modules have been developed, the responsibility then passes to the front-office units which, in addition to supplying information, participate in the design of the offer by assembling the modules designed by the upstream units and, thus, realizing the systems of products and services that most reflect the needs of the customers who are following. Often these units have the faculty to locally develop/modify some of the modules to satisfy the specific needs of the client (Cenamor et al. 2017), thus realizing a dynamic and partially decentralized design. If, up until a few years ago, this adaptation of the offer mainly concerned the service modules, lately, thanks to the growing diffusion of additive manufacturing tools such as 3D printers, the possibility of customization is spreading among the PSS-focused companies also for the product modules, which opens up a whole spectrum of new opportunities for creating value.

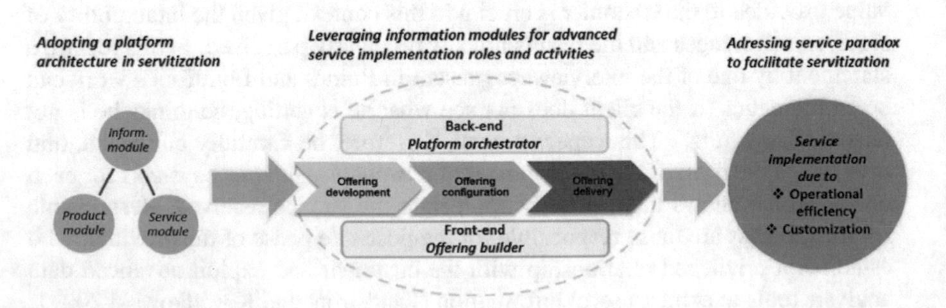

Fig. 6.7 The digital platform promotes the circulation of information, allowing back-end and front-end units to work together to create the offer (Cenamor et al. 2017)

The last element of the organizational structure that, with the adoption of a PSS offer, goes through important transformations is the network of actors that compete together with the provider for its development, or the so-called value chain.

In this context, there is an expansion of the value chain, which sees the entry of many new interpreters with a strong integration characterized by an intense activity of sharing information and resources.

The reasons for this type of choice are many and dictated by necessity. First of all, no company, even large and structured, is able to sustain the load of complexity and risks that come from the management of a PSS proposal (Gao et al. 2011) and, in order to face this criticality, many companies therefore opt for specialization limiting themselves to focusing on the core activities outsourcing the remaining processes to third-party companies.

It is then essential to establish with these network actors (suppliers of goods, service providers and distributors) deep relationships, which allow the creation of an integrated and organic proposal supporting the main provider in various activities such as designing new products/services, production planning, quality supervision and inventory control; in this way, despite a decentralization of supply management, its homogeneity is in any case guaranteed and indeed, being carried out by various specialized entities integrated with each other, the total value of the offer is higher than before. Concretely, this integration is often realized with common digital platforms, which facilitate the circulation of information and a continuous interaction between the parties in real time (Reim et al. 2015).

Degree of Servitization

Understanding how much servitization impacts on a company business model, and how consistent is the role of PSS in this *service-driven transformation*, is closely related to the degree of servitization that a company is willing to achieve. According to Martinez et al. (2010), four criteria can be identified to determine the level of servitization and, accordingly, how greater is the impact of the transformation. These criteria, which are reported in Table 6.1, must be considered with reference to the product–service continuum described in Chap. 1.

First of all, the "value basis of activities" refers to the value delivered to customers, considered as the primary driver to ensure customer retention and reiterated relationships over time: in the context of a high servitization level, value is perceived so as to be maximized over time thanks to long-term relationships, rather than on the basis of a single transaction.

The second criterion looks at assets: a high degree of servitization demands for a greater focus towards assets utilization rather than their ownership.

The "offering type" is linked to the well-known categories of PSS (Tukker 2004): whether a low level of servitization implies add-on of services to the offering, a higher degree of service-driven transformation is aimed at providing a consistent "total service integration" with a personalized solution for each customer (or category of customers).

Table 6.1 Criteria for the identification of an organization's servitization level (Martinez et al. 2010)

Criteria	High servitization	Low servitization	References
Value basis of activities	Relationship based	Transactional based	Gundlach and Murphy (1993) and Lambert et al. (1996)
Primary role of assets	Asset utilization	Asset ownership	Tukker (2004)
Offering type	Total service integration	Physical product plus extra services	Boyer et al. (2003)
Production strategy	Mass customization	Mass production	Gilmore and Pine (1997)

The last element, "production strategy", oscillates between the two extremes of mass production (low servitization) to mass customization (high servitization).

When an organization decides to enter the road to servitization and to develop a PSS offer, this choice impacts on various aspects of the business reality, modifying them in depth and involving a real revolution, which will be even more accentuated and necessary the more advanced and complex are the services that it intends to provide its customers.

Pre-sale and After-Sale Value Communication

The areas that face a considerable revolution in a company that decides to adopt a PSS proposal are those of pre-sale and after sale, which we could combine with the company's delivery system nomenclature. Since, as previously noted, the value of a PSS proposal is largely co-created, it becomes crucial to establish a relationship of trust and cooperation with customers favouring an intense relationship with a large exchange of data and information. The main human resources involved in these functions will be the ones responsible for managing this link. Given the great difference between proposing a product-centric offer and a centric service, they now need more extensive and transversal skills than before (Ulaga and Reinartz 2011). First of all, the profile suitable for this role must be equipped with some soft relational skills, such as the ability to empathize with the people creating with them sincere and solid relationships based on honesty and mutual respect (Baines and Lightfoot 2013); this competence, although it may appear abstract and of little competitive value, can be instead crucial in a servitized environment based on long-term relationships. Another element of disruption compared to an ecosystem based on the sale of physical products is the heterogeneous knowledge required to the personnel who must cover this role: technical and economic, thus ranging from the notions of business, managerial skills, the ability to understand the customer's operations and, finally, be endowed with a profound mastery of the range of services that company can provide (Kindström and Kowalkowski 2009). Without these multifaceted characteristics, a sales manager would be unable to fully understand the structure and dynamics of the company with which it is interfacing and, consequently, to support the customer in the development of an offer that reflects in the best possible way its needs.

Further competence required is, therefore, the ability to communicate the value provided in the pre-sale and after-sale phases (Reim et al. 2015). Given the intangibility of services, it may be difficult to make the customer aware of the benefits guaranteed and for which it pays. Concretely, it is fundamental to know how to develop adequate measurable parameters, but not limited to economic and performance parameters, but also of different nature since the benefits of a PSS offer are not only merely operational, but may also have wider scope (cohesion between companies, trust, attractiveness, etc.).

Short-Term and Long-Term Commitment and Retention of Customers
Final element that characterizes a PSS-related business model concerns the role of customers in the overall PSS offering, as already highlighted with the key element of value co-creation.

This latter factor also influences the contracts that regulate the relationships between the two actors a tangible transformation is the contractual structure, which is now much more complex and precise in defining what the established services are and that must be achieved by the provider's PSS (Reim et al. 2015) and the distribution of risks and responsibilities of the various processes that involve both parties. Also prices and payment methods of the various services and products obtained by the customer are established and calculated, not with reactive dynamics such as cost-driven pricing, but rather with the proactive value-driven mechanics; in many situations, this has involved a strong revision of the accounting mechanics, and the development of a new pricing discipline, which allows a model of elastic and dynamic revenue but at the same time allowing the company to cope with the new fixed or hidden costs that the provision of services implies for its budget.

The new complexity of the contract, the great dynamism inherent in the services and the precision required to determine the performance and distribution of responsibilities also heavily affects the number of actors involved in the bargaining process. While, in the traditional dynamic, is established a dual dialogue between the provider's sales function and that of the customer's purchases, with a superficial involvement of other functions, the offer is well-defined, with the adoption of a service, the context changes and evolves in a radical way, as already pointed out above in the paragraph. Now the offer presents nuanced details, with various types of services available and its characteristics are, then, outlined and agreed through a complex process of cooperation, which obliges interaction to various functions and actors of both value chain of the provider's client (Ulaga and Reinartz 2011).

Once the contract has been stipulated and all the elements that will outline the relationship between the two companies, start the so-called after-sale phase that, as already underlined, now assumes a much more central role than before: the service package established in negotiation phase must usually be provided for a very long period, sometimes even for the entire life of the product, which implies various changes in the company's value chain. In addition there is the need to develop tools and approaches suitable to keep the customer aware of the value supplied. This is a topic already analysed in depth and, certainly, one of the elements of greatest break with respect to the traditional delivery ecosystem. So, it is fundamental a capillary network of facilities distributed close to clients, with the task of providing the services requested by the latter and keeping with it a concrete and continuous relationship (Baines and Lightfoot 2013). This condition, which implies strong investments in assets and a distribution of resources on a global scale, is essential, as it allows to maintain contact with customers and thus "feel the pulse", which instead would not be as effective if relying on third-party service providers and on the filtered knowledge.

6.3 Unveiling the Key Elements of PSS-Based Business Models

This paragraph aim is that of presenting relevant cases of PSS implementation, so as to demonstrate how the concepts presented throughout the book can be put into practice. Table 6.2 summarizes the cases highlighting the relevant elements of the PSS business model that distinguish each case presented.

The boxes below report the first two cases. These are two interesting examples concerning the development of two different use-oriented offerings. The second one is an innovative case from the B2B context. These examples highlight distinctive elements and characteristics in the implementation of business models related to the use-oriented formula.

More specifically, the first case exemplifies the importance of two business model elements, which are the degree of servitization and the design of the offering, while Case 2 exemplifies value co-creation and design of the offering.

Table 6.2 PSS case studies and key elements of PSS-based business model

Case	Title	Design of the offering	Value co-creation	Functional integration	Degree of servitization	Value communication	Customer retention
1	Bike-sharing	X			X		
2	Co-working spaces	X	X				
3	Digitalization of solutions	X		X	X		X
4	IKEA	X			X	X	
5	Rolls-Royce			X	X		X
6	Circular strategy: eStoks		X	X		X	
7	Uber	X	X	X			
8	Airbnb	X	X	X			

Case Study

Use-Oriented, Same Category for Different Types of Success

Case 1: bike-sharing

Operating in out-of-home advertising and marketing sector, the firm launched the PSS offering in order to expand in new directions for advertisement, exploiting the high success and popularity of alternative means of transport, based on the concepts of **sharing and environmental sustainability**. This PSS brings to the firm an important competitive advantage and source of extra revenues, which though are mainly attributable to advertisement chances offered by the **bike-sharing business model**. Indeed, PSS is not considered as a "per se" offering, but is only linked to advertisement possibilities. Expanding in this new sector requires an important effort to retrieve **new capabilities and resources** and reorganize **internal/external processes**. Although capabilities have been acknowledged as a valid investment, they are not considered a key element of PSS's success and they are easy to retrieve. Conversely, resources and processes play an important role in this PSS offering. The formers are closely related to **investments in R&D** to develop **hardware and software components**; the latter consists of an entire series of **new organizational processes developed** for bike-sharing PSS that positively affected also the other activities of the company. Both these elements represent a **hard obstacle to replication** by competitors. The implementation of the bike-sharing offering has been **strongly affected by "success story"**. The introduction of the bike-sharing system in Italy and other European countries followed the first successful implementation in Norway: this experience raised the commitment to the model inside the organization and, more important, affected the acceptance from customers in the particular context of bike sharing. Indeed, the success of a bike-sharing offering can vary according to the place where it is implemented: in the U.S., mainly because of a different cultural background towards bicycles as a mean of transport, the answer from the market was very limited, while in Europe there has been a more enthusiastic response from customers. **Implementing a successful bike-sharing model is quite a hard challenge** because of the **high level of costs** and problems linked to the organizational and managerial aspects of the overall system. Thanks to the successful implementation in Norway (and subsequently in other Northern European countries) since the end of 1990s, the company could adopt the same model in many other countries and cities.

Case 2: co-working spaces

The use-oriented model employed by the company is focused on the **management of co-working spaces** for ICT start-ups. Start-ups can share **common places to work in a collaborative environment** and benefit from the presence of other similar companies, or otherwise there is the chance to rent private spaces like offices and meeting rooms and the **payment formula is based on**

monthly fees, which gives also access to clients to the entire network of spaces located in different Italian and European cities. The main reason behind the development of this business model is the exploitation of possibilities linked to the concept of co-working and sharing working spaces, so **clients do not have to face fixed costs** at the beginning of their activity while also benefitting from positive effects and influences deriving by working (literally next to other similar start-ups). **The presence of a network of spaces available** for customers **gives the company a privileged position over its competitors** and, at the same time, it **ensures a considerable flow of revenues and gains**. **Physical resources** mainly consist of spaces and settings, which can benefit of a distinctive design, while **capabilities** behind the overall organization of PSS offering are not distinctive, even if they constitute quite an important **intellectual capital and knowledge asset** of the firm. **The real distinctive element** behind the success and competitive advantage of the considered PSS are **organizational processes** and the presence of a network.

(*Information and data presented are taken from interviews conducted by the authors*).

Even if the two cases presented are very different and involve different types of offering in different contexts and markets, they tell two very similar stories of enthusiastic success of two innovative offerings. They highlight how important is the role played by intangible assets and tacit knowledge, and exemplify how companies can exploit chances offered by these distinctive elements in PSS design and implementation.

The next case demonstrates how soft skill development and interaction with customers are key elements of a successful PSS implementation as well. In this situation, there has been no particular infrastructure development, and the case shows how different can be PSS contexts, given the chance of implementing it as an entirely new offering, or an upgrade of existing business models. Furthermore, according to the specific situation considered, the impact on business models areas can vary for each example considered. Key elements of the servitization business model here considered are the degree of servitization, design of the offering, functional integration with partners and commitment/retention of customers.

Case Study

The Digitalization of Solutions

For the company, digitalization is not a phenomenon of recent implementation. Indeed, it had a specialization **in the field of Operational Technologies (OT)**, and in the **development of monitoring and control systems** for different types of assets. Many business units have long been implementing advanced sensors in their products, in some cases connected to a cloud infrastructure, or regulated by systems interfacing with each other, thus offering both traditional and digital services. What was missing, however, was a global homogeneity since **each business unit** in some way **managed its own digital processes in a separate way and was structured in a different way**. The company began a standardization process involving its various processes to define common guidelines for all its business units. Concretely, **this standardization took shape in the company digital platform (named Ability)**, to which at present more than 210 products/solutions/services are available for its customers. The industrialization of services allowed the company to formulate **advanced proposals, comprising the development of collaborative tailored solutions** for customers.

For the development of this platform the company, unlike its competitors, opted for a drastic choice: to build the platform it decided to **make use of collaborations, building strategic partnerships to complement its capabilities and know-how, gained through its experience in the OT world**.

Even a consolidated reality like the company considered, to succeed in developing its technological platform preferred to distribute risks and responsibilities with some strategic partners.

The digital platform, built on the Microsoft Azure cloud structure and enhanced by the partnerships with major ICT companies, thus **represents a potential source of innovation and value for customers**, and with its ability to interconnect with other existing systems also provides considerable flexibility and dynamism. Despite this, **there are some barriers** that are somehow hindering the diffusion of the company's digital solutions. These are basically three: **The first one is customers' resistance to change, who often are not aware of the potential of these innovative solutions**: to overcome this obstacle, **the company is improving communication and building awareness** of the advantages these new PSSs bring with them. **The second one, is that often customers are conditioned by not having the financial resources** needed **to start a digitalization process**.

The last barrier regards the skills required to implement these new technologies, which currently are not widespread: in this light, however, the company can support its customers with its know-how in the digitalization process in a collaborative perspective.

Moreover, another element of interest concerns the platform's name, Ability, representing one of its foundational concepts, **at the centre of the offer there is no particular technology but the man**, the specific set of **skills and experience** built by the company and its employees, which is now **capitalized and transformed into potential value for the customer** through a close collaboration that finds its manifestation in the provision of the service. Technology is regarded as a tool, an enabler for existing skills thanks to technology, skills can now emerge and be exploited in an optimal way, enhancing competition in a fast-changing environment. Concretely, **this led to a renewal and evolution of many services that the company provided**.

Finally, for what concerns the sales process of the company the sales employees **were largely provided with an adequate technical background** to effectively interact with customers. Therefore, **the transition to a standardized and more digitalized offer did not require any particular change** for the sales units' employees, **but rather changed the approach in interacting with customers**. Once again, **the focus shifts towards skills**: the sales process is no longer limited to a simple comparison of its technological offer with those of competitors, but the dialogue rather winds around making the customer understand what is at the core of the company's ability, its domain expertise and its know-how and how these characteristics, combined with the customer' own characteristics, can respond to specific needs that are a source of added value. The **post-sales phase** that follows is **a natural prosecution of this process of virtuous cooperation**.

(*Information and data presented are taken from interviews conducted by the authors*).

From cases presented above, PSS development and implementation can also appear more "simple". The case below presents how a simple PSS offering, related to the product-oriented category, might bring non-negligible benefits also to a big company with an established and robust business like IKEA. In this case, key elements of the business model are the degree of servitization, design of the offering, pre-sale and after-sale value communication.

Case Study

IKEA and the Strategy of Cost Leadership

IKEA is a multinational company founded in Sweden and specialized in the **sale of furniture**, furnishing **accessories** and other **objects dedicated to the home**. Born from a small village in southern Sweden, company has a special attention to quality while maintaining a low price. **The strategy is focused on reaching cost leadership**. Kampard, entrepreneur and founder of the company, noted that the well-designed products seemed to be destined only to the highest social groups, so he decided to offer a wide range of furniture of good design and functionality accessible to the majority of people. **To be able to meet various needs at the same time**, the **price tag is "designed" before the true idea of the furniture is born**. With 345 stores in 42 different countries around the world, the company has become globally famous and is able to attract a large number of customers each year. **Despite the limitations imposed by price strategy**, in recent years **IKEA has been increasing its concern towards the environment by promoting a servitization policy** to better meet the needs of an increasingly demanding customer base that is no longer satisfied only by low prices. The idea of IKEA has grown over the years by **combining social values** (reaching most people), **environmental values** and **value of use for the customer**. As already highlighted above, it is now impossible for companies to maintain a good position in the market and continue to compete on the price strategy.

IKEA, to build a stronger competitive advantage, decided to offer a "service experience" for its customers through the co-creation of solutions in a pre-purchase phase, to make the customer more involved and more confident in making the right choice. In addition to this, IKEA showrooms can be seen as "experience rooms" where they try to emotionally involve customers who have the opportunity to receive a real experience before the actual purchase.

The company's website provides an overview of the services that are currently offered, a series of **basic services to support** the simple **sale of the product which is given below**:

- **Design, measurement services and consultancy**;
- **Product availability and order status check**;
- **Transportation at home**;
- **Availability of pickup points**;
- **Assembly at home**;
- **Waste disposal**;
- **Product return and guarantees**;
- **Online assembly instructions**;
- **Parts replacement**.

Therefore, IKEA's offering can be seen as a basic type of PSS that **could be included in the product-oriented category**, where the ownership of the good passes into the hands of the customer at the time of purchase with the possibility of supporting services to complete the offer. Before the actual purchase, IKEA offers a basic online configuration service of its "room" based on style and size, immediately proposing a design of the solution with the corresponding estimate. Alternatively, **it offers the possibility to book an appointment with a consultant** to plan your home in more detail. The company's **cost leadership policy does not allow the creation of products** designed according **to the client's wishes** but, at the same time, **tries to combine the different standardized products** in order **to obtain the final solution** that best meets the different **needs of the consumer**.

In recent years, **IKEA has also undertaken customer services that go hand in hand with a service to the final consumer and an environmental benefit**. It should not be forgotten how **the concept of product service system binds with** a double thread to that of **environmental sustainability as it proposes, with the collection and reuse of used products, to reduce waste**

and the consequent environmental pollution. Companies operating in the market segment of cost leadership can hardly follow the product throughout its life cycle until complete reuse and disposal, as it would entail difficult costs to sustain and then maintain truly competitive prices. To be able to match the various ideas **IKEA proposes a disposal service both in the shop and at home of old furniture when a new product is purchased in store or at the time of delivery**, with the chance also to book **a service for disassembly and disposal for major structures** such as bathrooms and kitchens. Therefore, the company offers this service when the customer still decides to buy a new piece of furniture from their company and is confident to obtain a benefit from this service.

(*Information and data presented are taken from the company's website ikea.com*).

The following example, presents a well-known case, which is still relevant nowadays and shows many interesting insights on PSS development and successful implementation. The business model is focused on the degree of servitization, functional integration with partners, commitment/retention of customers.

Case Study

Rolls-Royce, The Strategic Importance of a Niche

Rolls-Royce Holding plc is **one of the most important examples** of success in the **integration between product and service** and has been particularly credited for being the first company that successfully launched contracts based on the final result. In the last two decades, in fact, the management transformed a loss-making British company into a world manufacturer of large jet engines by eliminating the difference between manufacturing and services offered. The company operated in a very competitive market and, for this reason, it was decided to develop new products with innovative ideas: from the use of carbon materials for the construction of the blades to the change of the basic architecture of the jet engines passing from the two three-axle shafts. The result was a more efficient product but, at the same time, more expensive and more difficult to design and build than those of the competitors. Rolls-Royce has also understood how, **in the field of engine production, the profits could be increased by proposing additional services** to the customer like the **maintenance and the sale of spare parts**. The first considered was an **incremental process** that aimed **at integrating** its **technology with the sale of a service to detach** itself more and **more from the competition** making the offer more and more inimitable. The key step that **led the company to significantly adopt the PSS business model** has been to move from the sale of a simple product to an **integrated service product**.

Already **in the '80s, the company introduced a "power by the hour" scheme**, where **customers had the opportunity to pay for engine maintenance as a fixed cost based on the hours of flight accomplished** exceeding a certain threshold. In this first moment, the customers still had to buy the product and, then, decide whether or not to activate the additional service. **The next step was the "Total Care" where each new engine had the ability to collect technical data and then transmit them to a control centre capable of storing and processing data**. This allowed the development of **intelligent analysis** to support and improve the ability **to predict engine behaviour and anticipate the need for spare parts and predict possible failures**. This solution allowed the company to improve post-sale efficiency and effectiveness, creating advantage for both suppliers and customers. To achieve these results, **the company relied on a large amount of internal and external resources** and, with the birth of the "Total Care" project, it needed a large amount of data and the ability to analyse them in real time with the consequent need for a development of skills within the company and a higher management capacity.

For the first time in this sector, Rolls-Royce, in fact, proposed **no longer the simple sale of the product** with the subsequent offer of spare parts and maintenance, **but the true functionality, the engine operation**. In this case, the company adopted **the types of PSS with the highest integration of the service**, the use-oriented PSS and the result-oriented PSS, typologies that require significant investments and changes in the entire corporate environment. **Rolls-Royce has, in fact, convinced its customers to pay a fee for every hour that the engine works, thus selling no longer the object but its functionality**; the customer may feel limited by no longer owning the product but, at the same time, it is guaranteed with a continuous maintenance and complete replacement in case of malfunctions or engine breakdowns. Rolls-Royce is offering services that last more than a decade and more than half of its engines in service are covered by this new type of contract.

The engines developed by the company are among the most advanced in the market to which are added a series of optional services of different sizes based on the precise requests of the customer. For example, in the field of aerospace engines, in addition to the aforementioned total care, it is possible to **choose between many different options** based on the need to have a **wide range of solutions or simply just certain areas of service** like **maintenance**, **efficiency control**, **resource management** and **after-sales assistance**.

Nowadays, an important fleet of operators located in the world allows to have a **close relationship with important customers** and, through the continuous monitoring of data, they are able to develop increasingly customized solutions with increasingly advanced technology. These projects would be unthinkable if the company would have continued to operate with the goal of reaching a high number of customers and not to design products of the highest quality, obviously at higher costs and therefore higher prices than competitors.

As regards marine engines, the company offers a wide range of services of total assistance. The "Customer Power Training" offer, even before the purchase, an experience on real engines or simulate a virtual reality. In the post-purchase phase, the company offers teams specialized in different areas able to solve different problems. Behind there is, therefore, an important investment on selected employees and training courses to allow staff to provide the best possible service.

(*Information and data presented are taken from the company's website rolls-royce.com*).

Throughout the book we stressed the importance of PSS even in relationship with "modern" business trends like, for instance, circular and sharing economy. The following boxes aims at presenting some relevant cases related to these two trends. The case presented in the first box is centred on value co-creation, functional integration with partners, pre-sale and after-sale value communication.

Case Study

Circular Strategy: Closing the Loop

The **boom in demand for electronic and electrical equipment** has led this sector to be **increasingly focused on activities such as manufacturing and disposal of end-of-life products**.

Two million tons of electronic material were added to the Brazilian market in 2012 and 1.4 million tons of electronic waste were generated, making Brazil the world's second largest producer of electronic waste behind the United States. Furthermore, only about 2% of this waste volume is reprocessed to be maintained in the production cycle.

The founders of eStoks realized that **about 5% of the products are returned to the manufacturer due to a defect or imperfection** and they can not be sold as new products due to the restrictions imposed by local regulations. These **discarded items constitute an untapped market worth £1.9 billion** contributing at the same time to losses and high volumes of electronic waste in Brazil.

The great vastness of the Brazilian territory creates logistics difficulties. In fact, most of the information technology producers are concentrated in the central and southern parts of the country, but a large part of the market demand lies in the north-east. **The creation of the value of this waste**, defined as high-value products with a defect, **is not feasible with a conventional reverse logistics model**. eStoks addressed these issues by creating a simple and smart approach to seize the untapped opportunity, while eliminating reverse logistics costs for electronic brands.

With a facility located in the north-eastern town of Recife, **eStoks collects products returned by local customers, replacing the original manufacturer**. After evaluating the status and quality of returned products, it defines the best strategy to ensure its usefulness and value.

Considering the volume, **50–55% of the recovered products are renewed** and **20–25% are repaired and resold**. **The remaining 10–15%**, consisting of the most damaged products, **is disassembled into parts**, and the components are **used for other repairs**.

eStoks, therefore, has the sales service of the regenerated products, chasing a **new customer segment**, at **cheaper prices**. In this way, high-quality technology and appliances are offered to a lower income audience.

The implementation of the circular economy has been successful. **Reconditioned products generate a value six times higher** than when they were recycled, access to high-quality products is provided at a very competitive price for low-income customers. There is a **reduction in logistics costs of up to 65%** and a future expansion is aimed at. In fact, considering this last point, the challenge of eStoks is now to promote and implement its services to more brands, manufacturers and retailers in the field of electronic and electrical equipment.

(*Information and data presented are taken from the Ellen MacArthur Foundation (ellenmacarthurfoundation.org/case-studies;* https://www.ellenmacarthurfoundation.org/case-studies/pre-consumer-waste-a-gbp-1-9-billion-opportunity-awaits) *and from the company's website estoks.com.br/*).

Similarly to the circular example, the following box will present some relevant cases related to sharing economy, related strategies and mechanisms to ensure the success of the overall PSS implementation. All the presented business models show as focal elements the value co-creation, the design of the offering and the functional integration with partners, with the last element playing a particular and relevant role in the context of sharing economy.

Case Study

Use Rather Than Own: The Sharing Strategy

Case 1: Uber

Uber is a company based in San Francisco (California) that provides a **private car transport service** through a mobile software application by connecting passengers and drivers directly. The company is currently show all over the world.

The platform was founded in 2009 by Travis Kalanick and Garrett Camp, only to be officially launched in 2010 in San Francisco.

The app **is changing the concept of mobility by putting drivers and passengers in direct contact**: people move with others who have the means. Uber provides a real service, so as to be included in the category of **companies that are part of the "on-demand economy"** that is "performs an economic activity created in digital markets and is able to meet consumer demand through immediate access to products and services".

The operation of the App is simple: users subscribe to the application by entering the data of a credit card and then, anywhere in the world where Uber is active, they can call a driver with a few taps in the display seeing the price of the trip. In addition to this, they know the route, the waiting time, the license plate number and the ratings on the driver made by previous passengers. At the end, it is mandatory to evaluate the trip. Cars can be booked by sending a text message or using the mobile application through which customers can also track in real time the position of the car booked.

In this entry, we consider all the entities that can come into contact with Uber's business: we, have on one hand, **the people who can interact with the company as customers or as drivers** (not as employees) and, on the other hand, everything concerning the external environment, which can be influenced by the actions of the organization.

A customer selects the company primarily to find an **economic alternative to the taxi**, while a person who decides to become an Uber driver essentially wants to make money, in many cases as a second job.

Uber has **two categories of customers: passengers and drivers**. Passengers are distinguished from those of a traditional taxi company because they must be equipped with smartphones and a credit or debit card. **The drivers are a segment of customers not present in the taxi market**. Uber uses freelance drivers, whose only requirement is to have a smartphone, a car in good condition, to be endowed with good looks and the desire to earn occasionally.

The automation and technology on which the company is based determine a fairly casual interaction with customers: the platform allows passengers to know perfectly where they will be picked up, without the need for further communications. The only point of contact is between Uber and the future drivers in order to illustrate the basic standards to be respected to offer an adequate service.

Several contact channels can be identified. Initially, the focus was shifting from one city to another, making sure that enough drivers and passengers had developed. Subsequently, the company has strengthened its position by developing the app and the website but, above all, the traditional "word of mouth" system, fundamental in a service like this.

Regarding the positive **value proposition**, we can identify the fact of **always ensuring a passage** and **having always available a passenger**, when a driver is free and wants to offer a ride.

Uber is also able to **measure demand in real time**, identify the moment in which it exceeds the available capacity and intervene by increasing prices, thus generating an increase in supply and bringing the balance back.

Another point is the chance of travelling without cash: using credit cards, the system is undoubtedly safer, for both parties. Moreover, passengers and drivers have the opportunity to know in real time the position of the other reducing anxieties and uncertainties.

Finally, the possibility for the customer to evaluate his driver with a rating (from 1 to 5 stars) increases the quality of service as those who have an average below a certain threshold are eliminated by the company.

For customers, critical points are security and privacy protection. Drivers are considered by Uber as freelancers and, therefore, not as company employees, so they do not have all the advantages of a regular worker.

The most important partners of Uber are the drivers who, owning their own car, allow the company to save the costs it would have to sustain by relying on a leasing company.

Another key partner is represented by local authorities, although at this moment not all of them are in favour of Uber's business model. The Uber PSS business model has three key pillars:

- the application is easy to use and practically free of defects;
- using the smartphone makes the service convenient and practical;
- there are no previous competitors in the taxi service and so the company established a solid infrastructure and a bond of trust towards consumers.

There are, however, **weaknesses in the Uber model**, in particular, regarding the **issue of insurance and legal battles**, due to **legal actions taken by taxi drivers** and unions in different cities.

Like other **partners**, there are also **investors** and **suppliers of the world mapping system**.

Main resources of Uber are:

- the **platform**, which allows drivers and passengers to interact with each other;
- the **pricing algorithms**, which are used to satisfy the value proposition, that is to guarantee the continuous balance between supply and demand of drivers and passengers and
- **routing algorithms**, which have the function of guaranteeing the minimum possible as regards waiting time.

The **key activities** of Uber are mainly two: **the development and optimization of the platform** to ensure its use and adoption by users and marketing activities with the aim of reducing the abandonment rate to a minimum.

The main objective of Uber is to consolidate its leading position in the city transport market as an alternative to the more expensive and less flexible taxi service. The company's **profit** is therefore **focused on the economic aspect**; despite this **there are some advantages both from a social and environmental point of view** (meeting with new people, fewer cars) but these are certainly not those that push people to use this service.

Uber uses its own servers to regulate taxation, using the phone's GPS technology to monitor all movements and charging the cost of passenger service based on miles travelled.

The biggest expenses for Uber are the development of the platform, hosting and salaries for IT engineers, sales team, marketing and various managers. Obviously, these are added to the salaries to be paid to the drivers.

Case 2: Airbnb

Founded in August 2008 and headquartered in San Francisco, **Airbnb is a portal on which people can publish, discover and book unique accommodations around the world**, either from their computer or from mobile phones or tablets. Whether it is an apartment for one night, a castle for a week or a villa for a month, Airbnb connects people through authentic travel experiences, at any price in over 34,000 cities and 191 countries. In addition, thanks to a customer service and a growing community of users, **Airbnb is the easiest way to earn from extra space available**.

Airbnb was born precisely in the autumn of 2007, and then officially announced in August 2008, with an idea by Brian Chesky and Joe Gebbia (the current CEO and CPO, respectively). Moving to San Francisco for the annual conference of the Industrial Design Society of America, and not having enough money to pay the rent, they decided to offer part of their apartment as accommodation to other travellers interested in the conference. It was simply three inflatable mats, from which the initial names Airbed and Breakfast were derived. So they created a very simple website and immediately three people booked for $80 each. This is how they perceived the potential of this simple, smart business that could offer great growth opportunities with minimal investment. In the spring of 2008, Brian and Joe decided to involve their roommate, Nathan Blecharczyck (still CTO), a Computer Science graduate at Harvard who had already worked in various positions for Microsoft, Opnet Technologies and Batiq.

In January 2009, the company progressed thanks to the intervention of Y Combinator, an incubator who invested money on this start-up. For 3 months, therefore, the company "moved" to Silicon Valley to work closely with the YC experts, to allow them to evaluate all its potential. Before the official presentation with the name Airbnb, the experimentation cycle culminated with a Demo Day, where the start-up presented itself to an audience carefully selected by invitation only. At this point, once the name has changed, the offer, which previously provided for the simple sharing of some spaces, widens its horizons to apartments, entire houses and any other type of property. In June 2010, the founders' loft became the company's office. In 2009, the company already had 15 employees and in 2010 registered 800% more bookings than the previous year, with circulation in 89 different countries. In 2011, it was considered one of the most important companies in America and obtained further funding from major investors such as Andreessen Horowitz, Digital Sky Technologies, General Catalyst Partners, Jeff Bezos, Ashton Kutcher. From this date onwards there has been an exponential increase from all points of view: Airbnb has wisely been able to merge the digital revolution of the 2000s with the request for simple accommodation at moderate prices and allowing people to create a community in it was increasingly social and global.

The ecosystem actors who come into contact with Airbnb are, on the one hand, all the **people who want to find a cheaper and "social" alternative to hotel accommodation**. On the other hand, there are those seeking for the possibility of **making available a room/house with the purpose of making a reasonable profit**.

The **needs of customers** who come to this service as guests are **both from an economic point of view**, as compared to Airbnb hotels is definitely advantageous, **both from the social point of view**, as this type of sharing of accommodation has become one of the points strong Airbnb, and therefore a reason for choice by those who want a different experience than the traditional hotel system. We find the **same types of needs** even by those who register on the platform **as a landlord**.

Airbnb has **three categories of customers**: **landlords/hosts**, **guests** and **freelance photographers**. **Hosts** are those who **make available unused spaces** and want to make a profit; to do this they can create a list on the portal, add data, useful information and set rental conditions. **Guests can choose**, according to their destination and availability on the site, **which spaces they would like to use**. Airbnb also owns a vast **network of "freelance" photographers** in all the major cities of the world who go to a place and **take photographs of the properties**.

Airbnb provides the customer with a **24/7 support service**. In addition, the service provides a **promotion and loyalty program**, so as to attract users for the first time. In addition to this, there is obviously the social platform of the service.

Positive value proposals affect both Airbnb customers. **For guests, using the App to find a room is quick and easy**, and the **rent costs are much lower** than those of a hotel. According to Brian Chesky, one of the co-founders of Airbnb, using the service is not simply renting a room but is receiving a sense of belonging to a community. Each type of content conceived on the site has been designed with the aim of creating a link and belonging among users; some examples are the guides to the neighbourhoods, real travel guides useful for those who need information about the city they are visiting, or videos co-created with users, made by travellers and assembled together in a single video of Airbnb. Another positive aspect for those looking for a rent is the personalization: by entering your preferences on the website you can find a room/apartment that meets your needs.

For the hosts, Airbnb offers the **opportunity to earn by renting their own room or the whole house/apartment**, as well as **offering them home insurance**. In addition, there is an evaluation system for both parties involved, useful for increasing the overall quality of the service.

The **main partners** of Airbnb, in addition to guests and landlords, are **regional real estate agencies**, who can rent their properties as hotels for extra income, **IT service providers** (web designers, hosting companies), **local photographers**, responsible for providing the unique aspects of each room/room through the Airbnb homepage, the regional government (for purchases of advertising space), **payment service providers, local cleaning agencies**. Finally, **investors**: Airbnb today has raised $4 billion through investments.

Airbnb has a **large network of hosts,** and therefore can offer a great choice to various customers looking for a rent. The **other resources** on which the company is based are all **web technology experts** who are able to keep active an easy-to-use website, **creative human capital**, that is, all people able to offer trendy and intuitive images of Airbnb through advertising and, finally, the **online payment system**, simple and quick.

The company's **key activities** are mainly those related to **advertising** (online and offline), **marketing**, **maintenance of web platforms** (both the site and the app), **customer relations** and various **sponsorships of local events**.

By using Airbnb and sharing accommodation with other people, it is possible to reduce both water and energy waste. In 2016 in North America and Europe, travellers who chose to travel with Airbnb instead of going to the hotel have helped to save an amount of energy equal to 900,000 homes, water equal to 10,800 Olympic swimming pools and reduced emissions equal to 1.8 million of cars.

Airbnb's main goal is to be able to expand its market leadership, exactly what Uber is trying to do in urban transport.

Airbnb retains a share from both landlords and guests: the first pay about 3% of the participation fee while guests generally pay between 6 and 12% of the reservation fee.

The **costs** that support Airbnb are **related to the payment of companies that deal with online payments**, the **creation and maintenance of online platforms** and the **insurance that the company provides to the landlords**. In addition, **the business is heavily dependent on human resources**: Airbnb needs to keep those highly creative talents to maintain its success. Finally, the company sustains **costs for heavy online and offline advertising campaigns** (e.g. billboards), as well as for event sponsorships.

The business model of Airbnb is very similar to that of Uber, in particular, it bases its fortunes on a technologically very high and intuitive platform, as well as on marketing and advertising very pushed.

(*Information and data presented are taken from companies' websites uber.com and airbnb.com*).

Seven Key Facts

- The product service system is a complete business model through which companies can put in action servitization strategies.
- The Business Model Canvas is a useful framework that allows to decompose a business model into nine building blocks.
- The business model process innovation framework can be adopted to support the process of business model design and implementation.
- Product service system brings in its definition a variety of elements that affect and transform traditional business models under different perspectives.
- Product service system imposed a shift in the focus of business models, moving from products to new concepts of service intelligence that go "beyond the product".
- A PSS-based business model is articulated in six key elements: Design of the offering, value co-creation, functional integration with partners, degree of servitization, pre-sale and after-sale value communication, short-term and long-term commitment and retention of customers.
- The key elements of PSS-based business model can interact in different ways to originate new successful offerings.

References

F. Adrodegari, N. Saccani, M. Perona, A. Agirregomezkorta, Business model innovation: a process model and toolset for servitizing industrial firms, in *Practices and Tools for Servitization—Managing Service Transition*, ed. by M. Kohtamaki, et al. (Palgrave MacMillan, Cham, 2018), pp. 309–321

R.H. Amit, C. Zott, *Business Model Innovation: Creating Value in Times of Change*. SSRN eLibrary (2010)

A. Annarelli, C. Battistella, F. Nonino, Product service system: a conceptual framework from a systematic review. J. Clean. Prod. **139**, 1011–1032 (2016)

C.M. Armstrong, K. Niinimaki, S. Kujala, E. Karell, C. Lang, Sustainable product-service systems for clothing: exploring consumer perceptions of consumption alternatives in Finland. J. Clean. Prod. **97**, 30–39 (2015)

T. Baines, H.W. Lightfoot, Servitization of the manufacturing firm. Int. J. Oper. Prod. Manag. **34**(1), 2–35 (2013)

T.S. Baines, H. Lightfoot, E. Steve, A. Neely, R. Greenough, J. Peppard, R. Roy, E. Shehab, A. Braganza, A. Tiwari, J. Alcock, J. Angus, M. Bastl, A. Cousens, P. Irving, M. Johnson, J. Kingston, H. Lockett, V. Martinez, P. Michele, D. Tranfield, I. Walton, H. Wilson, State-of-the-art in product-service systems. Proc. Inst. Mech. Eng. Part B J. Eng. Manuf. **221**(1), 1543–1552 (2007)

A.P.B. Barquet, M.G. de Oliveira, C.R. Amigo, V.P. Cunha, H. Rozenfeld, Employing the business model concept to support the adoption of product-service systems (PSS). Ind. Mark. Manage. **42**(5), 693–704 (2013)

K. Boyer, R. Hallowell, A. Roth, E-services: operating strategy—a case study and a method for analyzing operational benefits. J. Oper. Manag. **20**(2), 175–188 (2003)

J. Cenamor, D. Rönnberg Sjödin, V. Parida, Adopting a platform approach in servitization: leveraging the value of digitalization. Int. J. Prod. Econ. **192**, 54–65 (2017)

H. Chesbrough, Business model innovation: opportunities and barriers. Long Range Plan. **43**(2–3), 354–363 (2010)

M. Cook, T.A. Bhamra, M. Lemon, The transfer and application of product service systems: from academia to UK manufacturing firms. J. Clean. Prod. **14**, 1455–1465 (2006)

J. Gao, Y. Yao, V.C.Y. Zhu, L. Sun, L. Lin, Service-oriented manufacturing: a new product pattern and manufacturing paradigm. J. Intell. Manuf. **22**(3), 435–446 (2011)

B.J. Gilmore, J.H. Pine, The four faces of mass-customization. Harvard Bus. Rev. **75**(1), 91–101 (1997).

C. Grönroos, P. Voima, Critical service logic: making sense of value creation and co-creation. J. Acad. Mark. Sci. **41**(2), 133–150 (2013)

G.T. Gundlach, P.E. Murphy, Ethical and legal foundations of relational marketing exchanges. J. Mark. **57**(4), 35–46 (1993).

D. Kindström, C. Kowalkowski, Development of industrial service offerings: a process framework. J. Serv. Manage. **20**(2), 156–172 (2009)

D. Kindström, C. Kowalkowski, Service innovation in productcentric firms: a multidimensional business model perspective. J. Bus. Ind. Market. **29**(2), 96–111 (2014)

D.M. Lambert, M.A. Emmelhainz, J.T. Gardenr, Developing and implementing supply chain partnerships. Int. J. Logist. Manag. **7**(2), 1–17 (1996)

J. Lee, M. AbuAli, Innovative product advanced service systems (I-PASS): methodology, tools, and applications for dominant service design. Int. J. Adv. Manuf. Technol. **52**, 1161–1173 (2011)

S. Lenka, V. Parida, J. Wincent, Digitalization capabilities as enablers of value co-creation in servitizing firms. Psychol. Mark. **34**(1), 92–100 (2017)

C. Lerch, M. Gotsch, Digitalized product-service systems in manufacturing firms: a case study analysis. Res. Technol. Manage. **58**(5), 45–52 (2015)

V. Martinez, M. Bastl, J. Kingston, S. Evans, Challenges in transforming manufacturing organisations into product-service providers. J. Manuf. Technol. Manage. **21**(4), 449–469 (2010)

O. Mont, Clarifying the concept of product-service system. J. Clean. Prod. **10**, 237–245 (2002)

A. Osterwalder, The business model ontology—a proposition in a design science approach. Ph.D. thesis (2004). Available at: http://www.hec.unil.ch/aosterwa/PhD/Osterwalder_PhD_BM_Ontology.pdf

A. Osterwalder, Y. Pigneur, *Business Model Generation: A Handbook for Visionaries, Game Changers, and Challengers* (Wiley, Hoboken, NJ, 2010)

M. E. Porter, J. E. Heppelmann, How smart, connected products are transforming competition, in *Harvard Business Review*, vol. 92 (2014)

W. Reim, V. Parida, D. Örtqvist, Product-service systems (PSS) business models and tactics—a systematic literature review. J. Cleaner Prod. **97**, 61–75

D.J. Teece, Business models, business strategy and innovation. Long Range Plan. **43**(2–3), 172–194 (2010)

A. Tukker, Eight types of product-service system: eight ways to sustainability? Experience from SusProNet. Bus. Strategy Environ. **13**, 246–260 (2004)

W. Ulaga, W.J. Reinartz, Hybrid offerings: how manufacturing firms combine goods and services successfully. J. Mark. **75**(6), 5–23 (2011)

R. Wise, P. Baumgartner, Go downstream: the new profit imperative in manufacturing. Harvard Bus. Rev. **77**(5), 133–141 (1999)

Glossary

Business model Canvas is a model developed to allow graphical representation and analysis of an entire business model, divided into nine key blocks: Key Partners, Key Resources, Key Activities, Value Proposition, Customer Relationships, Channels, Customer Segments, Cost Structure, Revenue Structure (Osterwalder and Pigneur 2010).

Circular economy has been defined as "an industrial economy that is restorative or regenerative by intention and design" (Ellen MacArthur Foundation 2013). The aim behind this concept is that of implementing practices with a specific sustainable focus on design, maintenance, repair, reuse, remanufacture, refurbish and recycle. The goal consists in minimizing inputs and wastes of resources, energy emissions and leakages, by closing/narrowing loops of material and energy, in contrast with the traditional linear approach of production, consumption and waste generation (Geissdorfer et al. 2017).

Collaborative consumption according to the definition provided by Botsman and Rogers in 2010, Collaborative Consumption is a cultural and economic model based on access to goods rather than their exclusive possession; through technology and peer to peer, the movement reinvents the traditional concepts of sharing, borrowing, trading, renting, donating and exchanging.

Competitive advantage as originally defined by Porter (1985), indicates an attribute, or a series of attributes, that allows a company to outperform its competitors.

Competitive strategy is an action plan, developed for the long term, devised to build a competitive advantage.

Cost leadership is a competitive strategy that aims at obtaining the lowest costs within a given sector, thanks to the lower price of the competitors the company can, therefore, attract a higher number of customers. In order for a company to achieve this strategy, it is necessary that the reduction of costs is accompanied by

A. Annarelli et al., *The Road to Servitization*,
https://doi.org/10.1007/978-3-030-12251-5

the maintenance of the product characteristics, which the customer considers essential, and that the cost advantage is based on elements difficult or too expensive to replicate for the competitors.

Differentiation strategy is a competitive strategy that aims to propose on the market a product or service with characteristics that make it unique and inimitable, thus preventing competitors from proposing the same final good. Sometimes, the differentiation consists in the offer of a product/service not yet available in the market; on the other hand, it can be seen simply as a different perception on the part of the client through proper marketing activity.

Digital-driven industries is a term to indicate the contexts in which the digital transformation of business (i.e. digitalization) has profoundly influenced the competitive nature of companies and of markets where they operate.

Digitalization is a term used to indicate the Digital Transformation of Business, intended as a process that concerns the adoption of digital technologies at different levels of an organization.

Economic value added (EVA) is a measure of the financial performance of a company, determined by deducting the cost of capital from the operating profit, with an adjustment for taxes. It is a synthetic and effective way to express the true economic profit of a company, and the ability of a company to generate profit and richness as well.

Extended product is a term developed by Hirsch and Eschenbacher (2000) and is further developed by Thoben et al. (2001) that indicate an offering where product's functionalities are enlarged and improved by the addition of extra services. It is one of the main synonyms of PSS presented in the literature, even if it resembles more the Product-Oriented category of PSS.

Functional product indicates "an integrated system comprising hardware and support services" (Alonso-Rasgado et al. 2004) where services are intended to provide support to product's functionalities, not only through maintenance but also decision-making and operations planning, remanufacture and education, with the aim of providing the client with a function rather than the product itself.

Functional sale indicates an offering model in which the product is not sold to the customer and there is a contract stipulation between the provider and customer/user. This contract resembles renting/leasing contract even if it is more advanced, since it is focused (as the term suggests) on providing functions rather than products or physical components. It is one of the PSS synonyms event if, like the concept of Extended Product, it is more similar to a PSS category, i.e. the Use-Oriented one.

Industrial PSS (IPS2) is a term that indicates a "knowledge intensive socio-technical system" (Meier et al. 2003, 2010) characterized by the integrated and determined activities of planning, developing, providing and use of products and services including also software components in B2B contexts.

Integrated solution is a synonym of PSS, introduced by Davies in 2004. It has been defined as a combination of products and services tailored on specific customer's needs. In this concept, the focus is more shifted toward the integration element that is capable of providing an effective competitive advantage, more than the simple combination of products and service.

Market segmentation is a practice that consists of dividing a market or a broad group of customers into smaller groups called segments, on the basis of some characteristics that determine a specific profile of customers grouped within a specific segment. These might be common interests, shared needs, similar lifestyles or demographic profiles.

Mass customization is a strategy that consists of "producing goods and services to meet individual customer's needs with near mass production efficiency" (Tseng and Jiao 2001). Value is created by the specific firm–customer interactions (in production–assembly phases) aimed at creating customized products, while maintaining the production and cost-efficiency levels that characterize mass production systems.

Net present value (NPV) is the actualised sum of present and future cash flows, and provides a synthetic value to represent the ability of a specific product/offering/strategy to create value on a time window that might span several years.

Niche strategy is a competitive strategy that, as the name itself suggests, focuses on a specific niche of customers. It can be a cost-oriented strategy aimed at serving a restricted circle of consumers by offering a product at the lowest price compared to other competitors; it can otherwise be aimed at differentiation and therefore at offering a product at a higher price but, at the same time, customized for a specific consumer standard.

Operations management concerns the design and control of production processes within a company, together with the design (and eventual redesign) of business operations. Operations management must ensure the effectiveness and efficiency of business operations: by managing the entire production system, it is essential to meet customer needs (effectiveness) while using as few resources as needed (efficiency).

Operations strategy is a strategic plan that details how resources will be allocated in order to meet strategic goals by supporting the production process and related infrastructures. It can be seen as the connection point between High-level Strategy and Operations Management.

Path dependence is a term used to indicate a phenomenon for which minor or apparently inconsequential advantages can exert important and non-reversible impacts on firms' set of decisions, like for instance resource allocation, and determine effects of strategic "lock-in" (Arthur 1990). Liebowitz and Margolis (1995) clearly identified three different types of path dependence, distinguished in first-, -second- and third-degree path dependence. The first form, namely first-degree path dependence, arises when this sensitive dependence causes no actual harm or undesirable outcome. Second-degree path dependence is strongly linked to the presence of imperfect information: indeed, companies sometimes make efficient decisions on the bases of available knowledge, which turn to be inefficient in retrospect. The third-degree path dependence, similarly with the second one, appears when the dependence from initial conditions puts the company on an inefficient path exhibiting lock-in effects, but in this case, there was enough information to recognize the inefficiency.

Post mass production paradigm is "a system of economic activity capable of encouraging and sustaining economic growth without depending on mass production and consumption" (Tomiyama 1997).

Product–service system is a business model focused toward the provision of a marketable set of products and services, designed to be economically, socially and environmentally sustainable, with the final aim of fulfilling customer's needs (Annarelli et al. 2016).

Product-oriented PSS is the first and simplest category of PSS among those proposed by Tukker (2004), where the focus is maintained on products' sale, with the provision of extra services seen as a "plus" in the offering.

Product–service continuum is an ideal representation of the countless chances offered by Servitization in combining products and services. The word "continuum" is used in this sense as the opposite of discrete, precisely to convey this concept. At one extreme, there are pure products, while at the opposite one, there are pure services, and in the middle, lies an infinite set of possible product–service offerings that can be developed and implemented, deriving from the integration of products and services.

Result-oriented PSS is the third and most radical category of PSS among those proposed by Tukker (2004), and is focused toward the final result provided by the product, as the term "Result" highlights. The producer/provider and the client agree upon a result/outcome/performance to be delivered, with few specifications on the modes of delivery.

Service paradox is a phenomenon highlighted by Gebauer et al. (2005), for which increasing servitization leads to an increase in revenues, but it does not always coincide with an increase in profits; as observed in numerous cases, the provision of services often implies an increase in fixed costs, which together with the poor scalability of servitization, can go to erode most of the profits, making the adoption of this business model in fact counterproductive.

Servitization is a movement from an outdated focus on exclusively goods or services toward integrated systems or bundles of them, with services playing an always more relevant role in the place of products (Vandermerwe and Rada 1988).

Servitization value correction coefficient (SVCC) is a quantitative estimation index intended to act as a reference and a support to decision-making in order to assess and forecast the strategic value deriving from non-monetary factors that exert a concrete and relevant influence in PSS implementation.

Sharing economy indicates an economic system where there is a sharing of assets and/or services between the users that are usually private individuals. These goods/services that are shared might be owned by one of the users or otherwise, they might be owned by a company that makes them available to different users simultaneously. Nowadays, the term is an umbrella that comprises a plurality of concepts to express several kinds of sharing modes.

Sustainability in the economic/business context, refers to the process of balancing changes to harmonize investments, institutional changes, technological development and resource exploitation, with the aim of maintaining and therefore improving the potential of meeting human needs and ambitions, not only in an economic meaning. Indeed, Sustainability as a threefold nature, since the term can be declined into an economic, environmental and social dimension, with different aims and outcomes to be pursued.

Sustainability driven industries is a term to indicate the contexts in which the concern toward environmental sustainability has profoundly influenced the competitive nature of companies and of markets where they operate.

Use-oriented PSS is the second category of PSS among those proposed by Tukker (2004), where attention shifts from selling the product to giving access to its usage: this gives access to the same product to different customers in a limited time span, according to different forms of renting and/or sharing.

Value co-creation refers to a specific process in which customers might actively participate in value creation together with companies, for instance by participating in the design phase of the offering, or also in the production/assembly/delivery/installation phases.

Index

A
Added value, 11, 36–38, 44, 47, 127, 132, 138, 147, 148, 153, 157, 192

B
Bullwhip effect, *see* Forrester effect
Business ecosystem, 87
Business model
  canvas, 175–177, 203
  element, 2, 188
  innovation, 175
  innovation process, 203
  strategy, 2, 16, 143
Business to Business (B2B), 9, 25
Business to Consumer (B2C), 15, 47

C
Centric approach, 182
Circular economy, 55, 57, 64, 92, 110, 112, 182, 197
Collaborative consumption, 55, 59, 114, 115, 182
Competences, 71, 72, 121, 134, 159, 170, 180, 186
Competitive
  advantage, 13, 15, 23, 39, 57, 70, 71, 87, 99–103, 106, 110–112, 117, 126, 135, 138, 176, 182, 189, 190
  context, 96, 135
  strategy, 95, 97, 98, 102
Computer-Aided Design (CAD), 147, 148
Computer-Aided Quality assurance (CAQ), 158
Context, *see* Competitive context
Contract, 3, 8, 25, 38, 42, 44, 49–51, 75, 109, 118, 124, 147, 150, 154, 162, 169, 187, 194
Cost
  leadership, 101–103, 106, 107, 118, 120, 194
  structure, 40, 127, 138, 177, 178
Customer
  interaction, 6, 45, 52, 182, 185
  loyalty, 107, 108, 110, 122, 134, 138
  need, 2, 4, 15, 18, 22, 25, 31, 32, 40, 73, 82, 98, 105, 120, 124, 151, 155, 180, 183, 184
  satisfaction, 18, 19, 118, 120, 176, 177
  segments, *see* Market segmentation
  value, 22, 31, 37, 72
Customization, 2, 20, 25, 36, 37, 51, 72, 89, 147, 169
Custom measurement systems, 183
Cycle of interaction, 46

D
Decision-making, 65, 125, 127, 133, 135, 136, 138, 151, 157
Design
  product, *see* Product design
  service, *see* Service design
Design for integrated solutions, 147
Differentiation, 14, 22, 57, 101, 104, 105, 107, 117, 136, 175
Digital-driven industries, 85
Digitalization, 24, 57, 87–89, 91
Dimensions of PSS design, 6, 146, 147

A. Annarelli et al., *The Road to Servitization*,
https://doi.org/10.1007/978-3-030-12251-5

E
Economic sustainability, 74
Economic value added, 127, 136
Ecosystem, 87, 90, 188, 201
Effectiveness, 8, 90, 107, 157, 183, 195
Efficiency, 11, 24, 34, 59, 65, 69, 86, 90, 102, 103, 106, 127, 157, 169
Energy saving companies, 66, 69
Environmental sustainability, 7, 77, 78, 175, 189, 193
ESCO, *see* Energy saving companies
EVA, *see* Economic value added
Extended product, 6, 14

F
Fast track total care design process, 147
Financial resources, 44, 161, 170
Fixed rate, 5
Forrester effect, 158
Framework, 5–8, 20, 41, 75, 78, 110, 113, 126, 127, 135, 177, 179, 203
Functional
  product, 6
  sale, 6

H
Heterogeneous concept modelling, 147, 149, 150
Human resources, 4, 12, 24, 38, 39, 42, 159, 170, 186, 202
Hybrid product, 9

I
Implementation, 5, 22, 32, 41, 46, 69–71, 74, 89, 93, 98, 99, 110, 112, 117, 133, 136, 151, 156, 159, 161, 175, 180, 188, 189, 194, 203
Industrial PSS, 9, 12
Industry, 11, 65, 66, 69, 93, 97, 102, 106
Information System, 2, 6, 146, 169
Inimitability, 110, 121
Innovation management, 88
Integrated solution, 9, 15, 37, 57
Integration, 16, 55, 91, 119, 144, 149, 151, 152, 155, 162, 163, 167, 182, 185, 186, 190, 194–196, 203
IPS2, *see* (Industrial PSS)
ISCL system, *see*Integrated service CAD with life-cycle simulator

K
Knowledge, 6, 14, 24, 33, 36, 44, 45, 47, 73, 88–91, 99, 100, 110, 117, 135, 159, 160, 170, 186, 190

L
Logical service, 145

M
Maintenance, 3–5, 18, 31, 32, 36–39, 42, 43, 46, 47, 57, 63, 64, 71, 75, 79, 81, 83, 89, 99, 100, 102, 105, 112, 116, 118, 121, 124, 132, 144, 146, 153, 157, 163, 164, 167, 176, 195, 202
Manufacturing, 4, 5, 9, 11, 12, 15, 18, 23
Market
  B2B, *see* Business to Business
  B2C, *see* Business to Consumer
  segmentation, 108, 110
Mass customization, 37, 186

N
Net Present Value (NPV), 110, 136
Niche, 107, 108
NPV, *see* Net present value

O
Offering, *see* Value offering
On-demand economy, 198
Operations
  management, 14, 143
  strategy, 107, 143, 171
Ownership, 2–5, 18, 31, 41–44, 50, 75, 82, 120, 153, 193

P
Partners, 23, 40, 41, 43, 61, 71, 75, 135, 154, 160, 161, 163, 167, 169, 181, 196, 199, 203
Path dependence, 98–100
Payment
  formula, 81, 189
  model, 5
Perceptual mechanism, 182
Performance
  measure, 18, 187
Post mass production paradigm, 9, 12, 15
Process delegation services, 36
Process support services, 36
Product
  design, 49, 57, 170
  development, 8, 14, 26, 37, 91, 161
  differentiation, 14, 104, 106, 111
Product-centric approach, 47, 146, 182, 186
Product-centric proposal, *see* Product-centric approach
Product life-cycle services, 35, 36
Product-oriented, 8, 26, 49, 50, 70, 75, 99, 117, 133, 162

Product–service
continuum, 3, 20, 185
offering, 9, 26, 31, 161
Product service system (PSS)
classification, 3, 12
design, 70, 95, 143, 144, 150, 170
implementation, 5, 22, 32, 41, 74, 123, 133, 135, 143, 156, 161, 170, 175, 180, 188, 190, 197
Profit, 8, 18, 33, 49, 84, 89, 107, 108, 110, 127, 133, 176, 194, 201
PSS, *see*Product service system
Pushed servitization, 25

Q
Quality Function Deployment (QFD), 149

R
Relationship, 6, 14, 18, 21, 23, 24, 33, 44, 46, 52, 55, 70, 90, 106–108, 136, 156, 157, 159, 161, 162, 167, 168, 170, 176, 183, 185–187, 196
Resources, 12, 40, 47, 50, 55, 61, 71, 88, 90, 95, 97, 100, 110, 113, 115, 116, 119, 121, 132, 151, 156, 161, 178, 182, 189, 199
Responsive mechanism, 183
Result-oriented PSS, 4, 26, 109, 113, 195
Reuse, 4, 8, 64, 110, 112, 113, 115, 175, 194
Revenue
flow, 126, 127, 130
structure, 109
Risk
avoidance, 124, 125
category, 135
evaluation, 138
management, 92
reduction, 38, 51, 124
retention, 124
sharing, 124, 125

S
Scope, 117, 149, 179, 182, 187
Seizing, 111
Service
design, 13, 23, 71, 146
development, 6, 8, 152
differentiation, 104–106, 110, 138
intelligence, 181, 186, 203
management, 118
paradox, 24, 26, 89
strategy, 107, 143
Service CAD, 147, 148
Service model/Service explorer, 75, 147, 148
Service-centric approach, 182
Service-centric proposal, *see* Service-centric approach
Servicification, 6, 9, 11, 16
Servitization
barriers, 22, 87
benefits, 6
degree, 4, 20, 26, 70, 78, 185, 188, 190, 192, 194, 203
drivers, 21, 89
strategy, 1, 16, 138
Value Correction Coefficient (SVCC), 136
Sharing
economy, 55, 69, 92
Social sustainability, 13
Socio technical system, 6, 7, 9
Strategic value, 8, 133
Strategy
formulation, 96, 97, 99
Supplier, 4, 5, 14, 25, 35, 37, 42, 45, 49, 50, 86, 112, 132, 151, 162, 169, 184
Supply network, 65, 66, 88
Sustainability
driven industries, 74
economic, *see* Economic sustainability
environmental, *see* Environmental sustainability
evaluation, 126
Social, *see* Social sustainability
SVCC, *see* Servitization Value Correction Coefficient

T
Technlology
design, 45
management, 146

U
Unique Selling Proposition (USP), 22
Use-oriented PSS, 3, 26, 69, 123, 195

V
Value
co-creation, 25, 31, 32, 39, 45, 51, 181–183, 188, 196, 197, 203
creation, 9, 11, 39, 120, 181, 182
offering, 4
Value communication mechanism, 183

W
Win-win strategy, 122

Printed in the United States
By Bookmasters